# DEVELOPMENTS IN RUBBER AND RUBBER COMPOSITES—2

## CONTENTS OF VOLUME 1

# DEVELOPMENTS IN RUBBER AND RUBBER COMPOSITES—2

*Edited by*

COLIN W. EVANS

M.Phil., F.P.R.I., F.I.R.I., A.R.T.C.S., F.Inst.Pet.

*Formerly Technical Director, Dunlop Limited, Gateshead, UK*
*Chairman, International Standards Organisation Committee, Rubber and Plastics Products*
*Chairman, Rubber Industry Standards Committee, British Standards Institution*

APPLIED SCIENCE PUBLISHERS
LONDON and NEW YORK

APPLIED SCIENCE PUBLISHERS LTD
Ripple Road, Barking, Essex, England

*Sole Distributor in the USA and Canada*
ELSEVIER SCIENCE PUBLISHING CO., INC.,
52 Vanderbilt Avenue, New York, NY 10017, USA

**British Library Cataloguing in Publication Data**

Developments in rubber and rubber composites.—(The developments series)
2
1. Rubber—Periodicals
2. Rubber goods—Periodicals
I. Title II. Series
678′.2′05 TS1890

ISBN 0-85334-173-7

WITH 17 TABLES AND 47 ILLUSTRATIONS



Photoset in Malta by Interprint Limited.
Printed in Great Britain by Galliard (Printers) Ltd, Great Yarmouth

# PREFACE

This volume, the second in the series dealing with rubber and rubber composites and associated topics, describes and updates further products and branches of the rubber industry, namely cables, footwear, sports goods and latex products, while other sections deal with hazards and health and safety at work advice and short fibre reinforced elastomeric composite materials. The theme of miscellaneous and unusual applications of rubber and rubber-like materials has also been continued. Each chapter is written by a well-known specialist in the field under discussion.

Mention is made, although obviously within classified limits, of current and up-to-date techniques in the various topics discussed and, where possible, a look is taken into the future.

Thanks are expressed to all the contributors and their companies for permission to publish and also for the various illustrations which are separately acknowledged in the text. Thanks are also expressed to my secretary Miss Patricia McLeman and to my wife for her usual jovial, happy and helpful advice and encouragement.

COLIN W. EVANS

To My Family

# CONTENTS

# LIST OF CONTRIBUTORS

A. R. Carter

*Shoe and Allied Trades Research Association, Satra House, Rockingham Road, Kettering, Northants NN16 9JH, UK*

T. Eccersley

*Chemicals and Textiles Division, Department of Trade, Ashdown House, 123 Victoria Street, London SW1E 6RB, UK*

C. W. Evans

*Formerly Technical Director, Dunlop Limited, Hydraulic Hose Division, Earlsway, Team Valley, Gateshead NE8 1YR, UK*

A. Ford

*Bayer UK Limited, Bayer House, Manchester Road, Broadheath, Altrincham WA14 5PF, UK*

R. C. Haines

*Dunlop Sports Company Limited, 97 Bridge Road, Horbury, Wakefield, West Yorkshire WF4 5NN, UK*

G. Lawson

*Rubber and Plastics Research Association of Great Britain, Shawbury, Shrewsbury, Shropshire SY4 4NR, UK*

R. E. Whittaker

*Shoe and Allied Trades Research Association, Satra House, Rockingham Road, Kettering, Northants NN16 9JH, UK*

# *Chapter* 1

# CABLES

C. W. Evans

*Formerly Technical Director, Dunlop Limited, Gateshead, UK*

*SUMMARY*

*The uses and types of cables have been divided into industry applications, and include oil, coal mining and quarrying, transport, shipbuilding, domestic and power distribution. Present and future applications within these parameters are discussed, together with materials, standards and specifications and the latest trends in manufacturing processes and engineering plant and equipment.*

## 1 INTRODUCTION

Like most other products, electric cables have to be designed to meet specific requirements to accommodate the special needs of the industry to which they are to be supplied. For the purposes of this chapter, current and future usages in these major industries will include:

(a) the oil industry;
(b) the coal-mining and quarrying industries;
(c) transport;
(d) shipbuilding;
(e) domestic and miscellaneous applications;
(f) power distribution.

Other areas include telecommunications, electronics and aerospace, but the very high degree of specialisation to which cables are used in these applications is considered to be too limited in interest, and they are therefore not discussed. This omission, however, is not intended to imply a lessening of these important usages, as these too are very important cable outlets.

Before the Second World War natural rubber was used to the full and was in plentiful supply, with grades of excellent electrical properties readily available. In the mid-1930s Duprene[1] became available and was used where oil resistance and good weathering and environmental properties were needed. This of course was the name given to the early neoprenes[1] which today are classified as 'CR' rubbers. During the war years, home-produced PVC came on stream and GR-S, the forerunner of SBR, became available, at first slowly and then in ever-increasing quantities. The use of elastomeric materials for cables in the immediate post-war years was rather slow, partly because of the presence of thermoplastics and also because of their non-availability.

Since then, however, and especially today, the future is very bright because of many new polymers and also because of new processing technologies, such as new types of continuous vulcanisation and the availability of polymers in powdered and particulate forms.

The use of electric cables in various industries will now be considered.

## 2 THE OIL INDUSTRY

### 2.1 Offshore Installations

The main areas of cable installations within the oil and natural gas industries at the present time and in the foreseeable future lie in the oil-drilling and exploration rigs in the offshore areas, together with the deep-sea production platforms. In these connections, it is not surprising that the cable industry has come to think of these floating monsters either as ships, or indeed as large ships sitting on the sea bed. There is one big difference, however, and that is that there is a greater fire risk on an oil platform than on an ordinary vessel, and because of this the need is for fire-resistant cables similar to those used by the navies of the world. Some of the designs for this still use PVC both for insulation and sheath. Additionally, fire retardance is considered to be a very important factor, with cable sheaths having high oxygen index ratings to ensure maximum resistance to flame propagation to other parts of the rig. Continuous research into compounds is an ever-present task.

In the event of a fire, sufficient time must be available to enable the workers in the vicinity to take to the boats and evacuate the premises, so anything that can be done to delay the spread of the fire is an obvious help and necessity.

European technology favours the use of EPR-insulated and CSP-sheathed cables for ships, and thus also on the drilling rigs, as these are considered to be ships for this purpose. Vertical runs of cable of up to 150 m can occur on occasion and it is because of this that sheath compounds are specified to have oxygen indices which can be even higher than 31, to limit flame propagation.

The early rigs installed in the North Sea were gas production platforms and were very small in comparison with today's and tomorrow's oil rigs. Indeed the facilities available on modern rigs can be likened to those of a small town's electrical requirements. Installations capable of 70 MW per platform are quite common. Therefore, depending upon the basic electrical design requirements, this can be distributed at British, American or European standard voltages between 3.3 kV and 15 kV, and it is now common practice for cables for these voltages to be used on modern platforms, together with the lower voltage distribution and control cables.

Medium voltage on American designed rigs is usually catered for by 5 kV crosslinked polyethylene insulated cables, but elsewhere EPR-insulated screened cables are generally used, irrespective of whether the main cabling is PVC or EPR insulated.[2]

Elastomeric cables are often preferred for their greater flexibility in situations where cables must be drawn into ducts after assembly of the platforms, and in some cases where their continuous temperature rating of 85 °C allows smaller conductor sizes to be used.

Historically, cables used on the early gas production rigs, where only relatively small power requirements were required for lights and small motors, etc., had mainly PVC only for both insulation and armoured sheathing in the 400–600 voltage range. The early oil production platforms which followed the gas exploration and production eras had electrical generation capabilities of several megawatts and rapidly developed in size to 70 MW.

It is usual for an oil production platform to function first as a drilling platform, and whilst the drilling equipment can be driven mechanically it is more frequently driven and controlled electrically via cables to the drill rig tower. These cables range from multiple small-core constructions up to single-core cables of 300 mm$^2$ conductor cross-section. Of necessity they have to be very robust and flexible to allow movement of the rig to new positions after the completion of each well.

Flexible armoured cables have been developed for this,[2] and these have been based in construction on mining cables. The flexible conductors are EPR insulated with very heavy-duty PCP sheath, armoured, and then a further PCP sheath. Additionally, as well as this application such cables also have widespread uses, ranging from reeling drum installations to helicopter starter leads.

Mention has been made earlier of the requirements and need for flame-retardant properties in cables used on oil rigs. This is to allow the cables to operate and function normally for an adequate time should a fire occur. Some cables have been designed which are capable of 20 min operation at 650 °C with no risk of conductor to conductor short-circuit.[2]

In the smaller size range these essential cables are usually insulated with silcone rubber, and for sizes upwards of 25 $mm^2$ cross-section, EPR-insulated CSP-sheathed braid-armoured cables to BS 6883 are quite suitable. The outer portion of CSP and armour gives adequate fire protection for a satisfactory period of time to the inner EPR insulation.

Silicone-insulated cables are also suitable in such hot areas as flare stacks, where they are used with ignition equipment and to navigation lights and other aids.

## 2.2 Submarine Cables

Additional to the offshore applications on the rigs themselves, submarine cables are used between the platforms in a group, as such a set-up offers an alternative to generation on individual platforms. Historically, paper-insulated and lead-covered cables were used, but modern technology has replaced these with elastomeric cables, despite the fact that paper insulation is unrivalled for high voltage and high current applications. Elastomeric insulated cables have proved satisfactory in areas as high as 84 kV AC, as they have the advantage of lighter weight because of the absence of lead, which in turn also permits longer lengths even up to 10 km in one piece.

One of the more difficult operations in very high voltage cable manufacture over the years has been the formation and elimination of voids at the insulation/screen interface. This produces ionisation discharges and ultimate cable breakdown. EPR insulation is ideal for submarine cable insulation because of its excellent resistance to ozone produced by partial discharges and electrochemical treeing. Triple extrusion techniques have been developed whereby both conductor and core screens are extruded in the same operation as the insulation.

The flexibility of submarine cables is an essential feature of their construction and the only positive way of achieving this is by the use of cables based on elastomeric materials. Such cables were used when articulated flare towers, pivoted on the sea bed and allowed to move under the influence of wind and wave, were introduced.[3] At that time (1975) the working stress under which EPR-insulated submarine cable insulations were required to operate was a maximum of 2·5–3·0 kV/mm with a maximum conductor temperature of 85 °C. However, because of this flexibility it is necessary during the laying of these cables on the sea bed to have a slight tension from the laying vessel in order to prevent loops being formed which ultimately cause kinking.

It is a fact that cable making and hose manufacture are very similar. This is certainly true with regard to the plant used,[4] which includes compound mixing,[4,5] extrusion, braiding, continuous vulcanisation and lead-covering processes and operations. It is therefore not surprising that an entirely new and up-to-the-minute development, of so-called 'umbilicals' or 'bundles' of hoses, cables, and combinations of both, has been developed for such applications, as the means of controlling the blow-out preventer stacks on drill ships and downhole logging of oil wells. The cables are modest in size and voltage, but have to be sufficiently robust to withstand extremes of temperature and pressure, and winds and seas of extreme violence. Hence they are armoured with high tensile strength steel wire.

The application of the cable-making technique of helical lay-up associated with submarine and other power cables ensures bundle flexibility without undue strain on the assembly, a vital factor in bundles used for blow-out preventer control purposes. This type of bundle normally consists of textile-reinforced thermoplastic-type hydraulic high-pressure hoses consolidated into a single unit, with a thermoplastic protective over-sheath, usually of PU. Such hoses have a very low volumetric expansion.

Umbilicals for production wellhead control may be electromechanical and are normally double wire armoured for ballast and mechanical protection purposes on the sea bed. Production wellhead control bundles can be manufactured in lengths up to 2000 m, depending upon size and number of hoses and cables.[6]

New handling and production techniques had to be developed in order to manufacture the bundles, especially the long length ones. Much larger spools of textile yarn had to be made and used so that the textile joins in the braid were fewer in number, i.e. further apart, and in this area also special splicing techniques had to be introduced in order to keep the braiding knots at very minimal levels. The textile portion of the hose reinforcement of such a high-pressure hose would normally be of high tensile steel-wire braid, but obviously as the environment is salt water this was not possible, so a specially prepared Aramid yarn was used.[7] Each length of hose has a lining of nylon with a cover of either nylon, polyurethane or PVC, depending on the exact operating conditions.

Prior to the introduction of the laid-up umbilical, single hoses were used and were tied together if more than one length was required, but this was far from satisfactory and was even hazardous in application, as loose hoses around an oil platform and floating in a North Sea swell is not good practice. In order to obtain a neat and compact laid-up bundle without twist, it was necessary to resort to cable technology and the laid-up hoses and cables were in turn covered with a thick sheath of PU, although other materials including PVC can be used as necessary.

**2.3 Shore-Based Installations**

In oil refineries the power requirements are very different from those used in general industry. The main difference is the possible presence of oil spillage into the ground, and any cables buried in such conditions have to be capable of withstanding such adverse conditions.[8] Even the best of oil-resistant polymers will swell to varying degrees and eventually allow oil to pass through and travel along the conductor to the terminations. Some plasticiser extraction is also fairly certain, and this can produce brittleness. Thus the cable will have only a limited life. The only sure way of preventing this oil seepage is to protect the cable cores with a metal barrier such as a lead or aluminium sheath.

Elsewhere the oil industry has parallel oil drilling and exploration work continuing on land as well as offshore, but this has not presented too great problems to the cable maker. The majority of cables in use have well-proven track records and are performing well, and this also applies to tanker terminals, refinery extensions and gas-processing plants. These are basically paper-insulated lead-covered cables, but EPR tends to be preferred for the voltage range 5–33 kV. Cables for lower voltage applications may be paper or PVC insulation with lead covering, but where oil and other hydrocarbon seepage is not present, PVC insulation and sheaths with armouring are increasing in use.

The rig and platform yards either use paper or are EPR insulated, the trend being for the latter type at high voltages with PVC-insulated and armoured cables for the lower voltage range.

However, the newer development of concrete structures has produced a more novel application. When the lower portion of a concrete structure has been completed in an excavated basin on the land, it is floated out for completion just offshore. The concrete pumps have to run continuously and the electrical linkage is maintained via an 11 kV EPR-insulated cable which can be either supported on a pontoon bridge between the structure and the shore, or along a line of buoys or by a submarine cable. This could well be a future area where the integral hose flotation system used for oil suction and discharge purposes for hoses could perhaps be built into and around the cable with the elimination of other flotation necessities.

## 3 COAL MINING AND QUARRYING

Cables used for mining and quarrying purposes have to be extremely robust with great flexibility, and additionally those used in underground mining have to have special fire-resistant properties as well as some conductivity or rather anti-static properties included in the sheathing compounds.

The source of electrical current in a mine is usually down a pit shaft and then along the roadway to the pit face. The cables in the pit shaft are cleated to the sides and are thus fixed permanently in a vertical position. They must resist water seepage and must not contain any ingredient that will flow out of the end of the cable. In order to resist occasional and external damage as well as to bear the strain of their own vertical weight in the pit shaft, double wire armouring is usually applied within the construction. A typical construction is one of plasticised PVC for installation with galvanised wire armouring impregnated with a no-flux bitumen and PVC outer sheathing. These cables operate at 3·3 kV, and after the shaft usually carry on along the roadway as far as the gate end, where the voltage is reduced to 550 V or 1·1 kV. As the pit face advances, of course the gate follows, so the 3·3 kV cable is lengthened from time to time. From the gate, pliable cable armouring is used and the final cables to and at the face are the trailer cables. The flexible cables supply power to the conveyors as well as for permanent lighting and battery recharging. Very often they are installed alongside and parallel to the coal conveyors, where apart from cut damage due to rocks and coal, etc., there is also the hazard of crushing, hence the need for flexible conductors, rubber insulation, flexible wire armouring and a very tough sheath of PCP.

Trailing cables also carry current from the end of the armoured cables to the cutting and drilling machinery at the actual coal face, and the operating conditions here are obviously very arduous, as not only do they have to withstand crushing but they have to be pulled and dragged around over rough rocks, etc., where they get bent and twisted into all kinds of unnatural contortions, and so have rather limited life cycles when compared with normal service conditions elsewhere. Short-circuits can occur due to conductor breakage and puncturing of the insulation, thus causing shorts and possible fire hazards due to arcing.

This is obviously very dangerous since if methane gas happens to be present, an explosion could occur. Therefore, on these cables each power core is individually screened either with a copper–nylon braid or more recently with conductive rubber. This must have a resistivity of less than $10\,\Omega^2/\text{cm}$, and hold this resistivity despite severe flexing, otherwise the circuit breakers will not operate and shut off the current.

The need for high tear and cutting strength is obvious in a coal mine, as is the need for fire resistance, and even no-afterglow requirements are now built into many mining specifications world-wide. This is to ensure that the cable will not re-ignite after the flame has been extinguished. In the future it is quite likely that even higher voltages will be used at the coal face, but the limiting factor on this is the availability of flameproof transformers and switchgear of adequate safety factor. The cable industry has already developed the insulation compounds and design features of such cables.

## 4 TRANSPORT

Much of the recent and foreseeable development work on cables for transport has been in the field of the railways, including underground trains. In this latter connection, considerable effort has been expended world-wide in an attempt to develop a so-called ‘smokeless cable’. This is a cable which when burned does not emit any smoke or toxic fumes. The smoke hazard is extremely important, not really because of its toxic danger, but because of the fear and panic it brings to the average traveller in a confined area such as a tunnel. In conventional PVC cables, during burning, copious black smoke together with a high level of hydrochloric acid gas is produced and this latter is a very nasty irritant to the respiratory system. Borates have been found to be effective in reducing smoke.[8]

The solving of this problem became urgent because many towns and cities world-wide have decided to install underground passenger travel systems as a solution to their traffic congestion and energy problems, and the underground transport engineers were faced with the possible problems of cable fires in the tunnels.

One way of minimising this is in the cable development[2] incorporating elastomeric compounds where the insulation can be flammable, but if burnt will produce neither dense smoke nor irritant gases. Over the laid-up cores a specially designed elastomeric filling compound having flame-retardant properties is applied. The function of this is to protect the insulation from external flame. The outer sheath is a flame-retardant silicone rubber. Very impressive results are being obtained with such cable constructions.

Another nuisance and possible hazard in underground train tunnels is the accumulation of brake dust and metal filings along the cables which are cleated to the tunnel walls, as this can catch fire owing to sparks or static. The fires then creep slowly along the cable unless checked. One way of stopping this was the use of asbestos braiding over the cable, but this is not now considered a healthy thing to do, owing to the possibility of asbestosis. However, materials other than asbestos can now be used in this connection.

Another large and very important area on the railways for cables is on the locomotive. This has occurred since the demise of steam in favour of diesel and electric traction locomotives. 'Diesel' locomotives contain more electric cables than may be realised. In fact, they should really be called diesel-electrical, since the diesel engine is used to drive dynamos and alternators to generate the power that is used for the final drive. Additionally, large air blowers are electrically driven to keep the engines cool and current is also required for the lighting and heating of the carriages. To make life difficult for the cable designers the cables also have to be resistant to diesel oil, and internal combustion engines are not particularly clean in operation, throwing oil about all over the place. One has only to look at the average diesel locomotive in use today anywhere in the world to see this. A railway diesel locomotive could almost be described as a small power station on wheels. Despite the cooling fans, operating temperatures in the confined space available in a railway diesel locomotive are high, and cables have to be rated for 85 °C operation. To date not all the required properties can be met at one and the same time, since elastomers are not available to cover the whole gamut of requirements simultaneously.

Ideally, according to British Rail,[9] their requirements are for a cable with the following properties:

1. Operating temperatures not less than 85 °C.
2. Low-temperature flexibility.
3. Non-sticky surface.
4. Oil resistance.
5. Non-flammability.
6. Insulation resistance to be as good as EPR/CSP.
7. Non-explosive pyrolysis products.
8. Non-toxic pyrolysis products.
9. Smoke production to be minimal.
10. Overall diameter to match existing stock of terminals.
11. Stranding to match existing stock of terminals.

Finally, with the advent of higher running speeds of trains the original mechanical signalling and points systems have had to give way to what became a major re-signalling programme using powered signalling boxes controlling 80–100 miles of track. The cables now have to operate colour light signals, points motors and indicate train positions and movements, often over long distances. Power attenuation is therefore a major problem and relatively large conductors are employed. Many of these cables are insulated with NR/PCP with PCP sheathing and are very suitable to meet conditions such as being dragged over ballast and damage from ballast machines as well as from weather, storm, rain and tempest, plus soaking in diesel oil and other oils on occasion.

In the realms of other forms of transport, shipbuilding is discussed separately below but, as stated as the outset, aerospace requirements are considered to be too specialised for discussion in the general light of this chapter.

## 5 SHIPBUILDING

A ship can be considered as a sea-going building when its requirements for electric cables are examined. The main differences, of course, are that a building is static whereas a ship is subjected to rolling and vibration movements because of the sea swells and wave motions and also transmitted vibration frequencies through its metal construction from the engines, etc. Conductors therefore have to be flexible and usually consist of strands of wire rather than single solid wires. Additionally a ship's travels produce changes in ambient temperatures, although apart from the general merchantmen which literally travel the globe, specialised ships such as banana boats, oil tankers and liquefied petroleum gas carriers tend to ply between a minimum number of ports. Because, in the main, ship's wiring is fixed and out of contact with human beings, its operating temperature has been up-rated to 85 °C, thus allowing smaller and lighter cables to be used. These consist mainly of EPR or XLPE cores with CSP or PCP sheaths.

Because the fire hazard on a vessel is far more serious than on land, fire resistance is very important, especially in emergency circuits where special fire-resistant cables are mandatory.

It is usually accepted that as cables are installed during the building of the vessel they have to last the life of the ship, otherwise a major rewiring operation becomes necessary. This prompts the question, what is the expected life of the cable? Most ships average 20 years, although recently more specialised ships remain serviceable only for about 10 years because they are either worn out or are economically outdated.

## 6 DOMESTIC AND MISCELLANEOUS APPLICATIONS

Over the years the older type of NR-based insulated and sheathed cables have given way to PVC, and this is now almost solely used in present-day and foreseeable future housing requirements. In the main, operating temperatures are well within the PVC limits of 70 °C. However, in boiler rooms and in the proximity of such devices where 70 °C can be exceeded, EPR- or butyl-insulated cables have replaced NR. Indeed, wherever there is a known area of heat, such as radiant heat bathroom bulbs, PVC is obviously not used, as temperatures of 120 °C can be reached. Even EPR is suspect under such operating conditions and silicone rubber has been in use for some time. Crosslinked ethyl vinyl acetate is growing in use for this application and is in common usage in West Germany.

For domestic appliances flexibility is required in a big way and here PVC vies with NR. PVC has some advantage in easy stripping and being bright in colour, but has temperature limitations of 70 °C. In the USA and in Europe, depending on the going price of NR, blends of NR and SBR are also widely used and the cost balance of these two polymers is closely watched.

Other items used domestically include power tools, and cables usually consist of PVC, NR or SBR insulation with similar sheaths, although industrial and garage applications are usually more severe than around the house and sheaths of PCP are used here, which of course has good tear abrasion and oil resistance.

The so-called unkinkable domestic flex is for some unknown reason unique to the UK. It consists of a two- or three-core stranded or bunched braided flexible construction with a thin bedding compound immediately under the braid and cords laid up in the interstices of the cores, thus keying the braid into the bedding compound and giving the cable its unique non-kink properties, so very obviously absent overseas.

Modern cables have to be designed for ever-increasing uses. High-rise flats, for example, require lift or elevator cables which do not twist and also stay in line, particularly at high speeds. They must not be unduly 'thrown' or they can become fouled up with the lift operating adjacent. Obviously the cables must be fire-resistant.

Advanced X-ray equipment requires small-sized flexible cables and has to operate at very high voltages. These requirements are contradictory and very delicate design balances have to be made. The voltages are of the order of 50–75 kV. Very flexible earth screens have to be incorporated into the two- or three-core flexible constructions for safety reasons.

Electric arc-welding has been gaining in application and special cables are required here. They have to be highly flexible but sufficiently light in weight to minimise drag and operator fatigue, and of course must be safe. Voltages are low with currents very high and heat is generated. Chlorosulphonated polyethylene and EPR rubbers are ideal for such applications.

## 7 POWER DISTRIBUTION

In most countries electricity is generated in a relatively small number of large generating stations, some nuclear, others conventional, and these are connected together by a grid system. In the UK these are organised by the CEGB (Central Electricity Generating Board). This electricity is transformed down at a number of secondary power stations, which are under the control of the area electricity boards. The voltage for this is usually 11 kV. Further voltage reduction to 415 V is carried out at substations which in turn supply electricity to a small area of several streets of a town or in rural areas to the small transformers attached to poles. There are also the cables which are buried in the streets of towns and which in turn are tapped to supply individual houses.

The grid system cables or rather wires are slung between pylons and are not insulated. The 11 kV and 415 V cables are normally insulated with oil-impregnated paper. PVC is not used because, being thermoplastic, it softens under overload conditions and would therefore cause electrical breakdown of the cable, especially at peak load periods. In the USA and Canada, for reasons of cost, the 11 kV cables are usually slung from poles rather than being buried. To say the least they are very unsightly! Under pressure from environmentalists, steps are now gradually being taken to bury them. Paper-impregnated insulation for these applications is being replaced for the future with crosslinked polyethylene and EP rubber, and these materials are quite satisfactory under wet and buried conditions. Modern 11 kV cable installations have to operate at considerable electrical stresses and because of this their design includes semiconductive polymeric layers, applied against the conductors. This has the effect of redistributing the stresses and reduces the possibility of electrical breakdowns.

The buried cables also contain galvanised outer wire armour over the laid-up cores, and this in turn is covered with anti-corrosion and anti-pest compounds. More recently PVC, PCP and CSP have been tried and used for this purpose.

Conventionally oil-filled cables have been used for high-voltage power transmission and have an excellent track record. There is, however, a growing interest in the possible use of elastomeric insulated cables as links in 132 kV transmission systems. As would be expected, these types of cables are difficult to manufacture and call for great care in handling and using the insulation materials in order to obtain good dispersion of the fillers. Indeed, such are the cleanliness requirements during extrusion processing that air-conditioned mixing rooms and extrusion areas are necessary.

## 8 TESTING AND SPECIFICATIONS

The type of test chosen to prove whether a new cable design is suitable or not depends very much on the type of cable insulant and construction in question. In the case of paper-insulated cables there are well-proven and established tests for typing as well as routine procedures, but the changing tendency to elastomeric materials requires a different approach.

One very important test requirement is that of measuring the magnitude of discharges within the insulation at working or higher voltages, both in the finished cable length from a routine angle and also in a finite sample's length when subjected to bending and heat-cycling tests. There are, however, pitfalls in relying too much on these measurements only. In the first place there is the problem of determining what level of discharge is permissible for a given dielectric material, noting that the energy of the discharge will also be a function of the stress and therefore must be related to the cable design. Secondly, there is a great difficulty when carrying out long length discharge testing to maintain an adequate test sensitivity and to reduce background noise to low levels. Finally, there is the serious criticism that the length of cable tested is normally about 100 m, whereas this is only a small portion of actual length of service.

The length of test sample is more critical for extruded elastomeric cables than paper-insulated cables, because the latter exhibit a much smaller scatter of results in breakdown testing. Thus the breakdown strength will be much more dependent on the length of cable under test or in service. Tests are made noting both stress failures and time failures at arbitrary stress levels and plotting a Weibull distribution.

It has been shown[2] that in such testing procedures, from the given parameter for a given dielectric material, derived from such test results, it is possible to calculate the probability of failure for a given operating stress at the time for any cable dimension, including length, made from the same dielectric material and by similar production technology.

Experimental work has confirmed that this technique is valid, at least for times ranging from one minute up to one year. For longer times the accuracy must decrease, but the technique still provides a very good guide to service life. This statistical method of treating life curves and the breakdown distribution has gained international acceptance.

It is very evident that because of all the widely used types of different cables some form of rationalisation is necessary, and indeed the cable industry has dealt admirably with this problem, not only at national level but also within Europe and internationally. The result of this widespread co-operation between the world's cable makers and users has been the issue of several international recommendations and standards under the auspices of IEC, which is the electrical equivalent of ISO.

The two bodies work fairly closely together and have basically similar aims. Thus the IEC's recommendations have been adopted nationally in the main with only minor changes, as has the EEC, except that the latter has tended to carry rationalisation even further. The EEC body is known as CENELEC.

Internationally those issued so far cover ship's cables and general cables, these latter being mainly for domestic and building purposes. Power distribution cables have also received attention, as have railway cables, and current discussions cover mining cables and underground railways. All this excellent work has resulted not only in standardised requirements but also in test methods and procedures, and is a noteworthy achievement.

## 9 MANUFACTURING TECHNIQUES

### 9.1 Continuous Vulcanisation

Cable-making processes have made many noteworthy advances over the past half-century, perhaps the most important one from a quantity production angle being that of steam continuous vulcanisation, which has almost certainly taken over from the old, rather cumbersome but nevertheless sound, manufacturing procedure known as longitudinal covering. Continuous vulcanisation enables extruded cables to be immediately vulcanised on their own in a tube containing high-pressure steam. This enabled high line speeds to be achieved and became so developed and sophisticated that very small and quite large cables could be cured by this technique. Steam tubes are in use in the horizontal position, or rather with a 3° slope, with catenaries and half catenaries to prevent cables touching the tube bottom at critical stages of cure, and even vertical tubes in tall towers. These latter are for the larger-sized cables.

However, despite their efficiency, steam tubes do have a rather serious drawback. Because of basic physics, steam pressures and temperatures are of course interdependent. Thus to obtain higher temperatures in order to obtain higher running speeds, increased steam pressures have to be used and at these higher pressures the cables become deformed and squashed.

The recent development of the so called PLCV[10] (pressurised liquid continuous vulcanisation) system (Fig. 1) has overcome this problem in that the heating medium is molten salt, the temperature of which can be controlled completely independently of the superimposed gas pressure. The justified claims of such a system include:

1. The vulcanising temperature is independent of system pressure.
2. The system pressure can be applied independently of the temperature.
3. The curing medium is a eutectic liquid salt which is commercially available throughout the world.

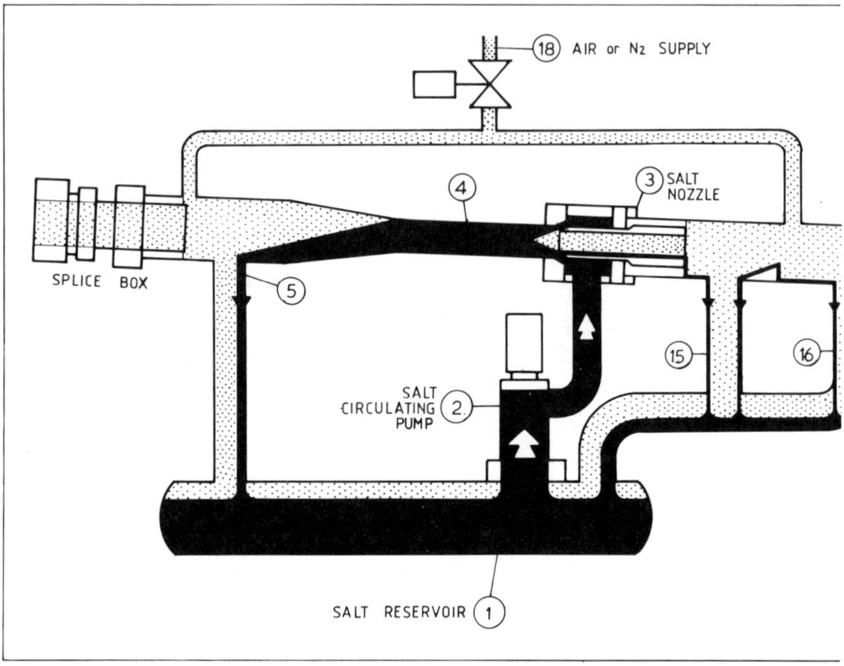


FIG. 1. PLCV system.

*Salt section:* The entire section containing the salt is heated to the required temperature by electrical trace heaters and is thermally insulated. The salt is stored in the reservoir (1), from which the special pump (2) takes a suction, delivering the molten salt to an injector assembly (3), the bore of which is the same dimension as the curing tube. At this point, the energy possessed by the salt is utilised to convey a body of salt up the curing tube (4) and to support the static head of salt which is in the curing tube, the tube being then completely full of salt. The salt exit (5) takes the circulating salt back to the reservoir (1). The injector assembly (3) acts as an entry point for the salt but, additionally, serves as a seal against back-flow of the salt, thus the cable does not pass through any mechanical seal on leaving the salt section.

*Cooling section:* The pressurised water-cooling section functions in a similar manner to the salt system, viz. water is stored in the water reservoir (6), the pump (7) taking a suction from (6) and discharging into the injector assembly (8),

4. The curing medium is cost-economic with a life counted in years.
5. The energy consumption for plant operation is considerably less than that required if steam is the curing medium.
6. The system makes possible continuous production of cables (and

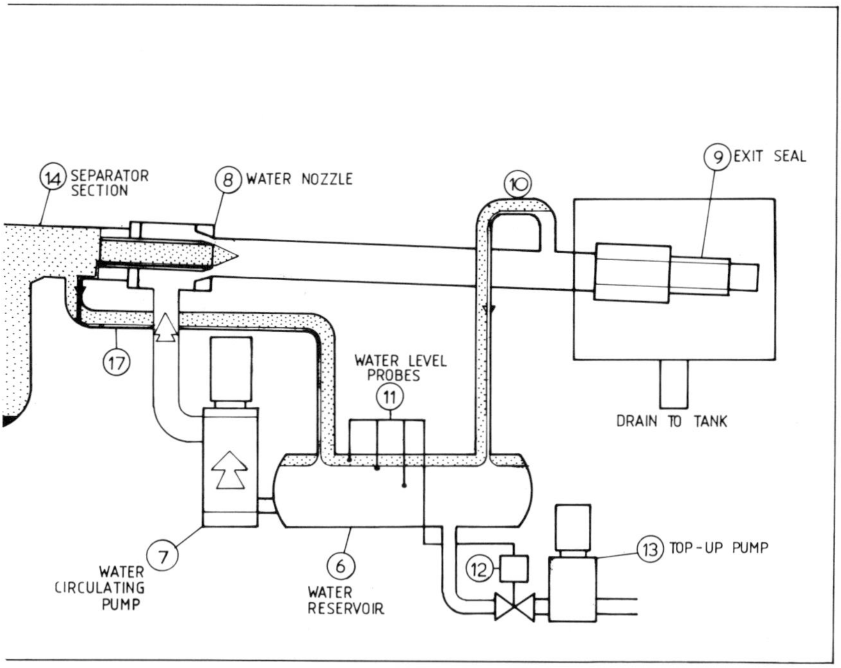


moving a column of water down towards the exit seal (9). Depending upon the degree of leakage from the exit seal, excess pump discharge is returned to the water reservoir (6) via the stand pipe (10). A simple level control (11) is located in the water reservoir (6) and a control valve (12) operates to permit the top-up pump (13) to maintain water in the reservoir to compensate for the exit seal losses.

*Separation chamber:* The separator section (14) is positioned between the salt and water sections and contains two return legs (15 and 16) to permit salt which falls from the cable after leaving the salt section to return to the salt reservoir. Additionally, on termination of production and stopping of the salt pump, the entire contents of the curing tube will return to the salt reservoir via legs (15 and 16). The third leg (17) permits any back-flow from the water cooling section to return to the water reservoir. The means of pressurising the system is provided by connection (18) from an external source of compressed gas. This connection (18) is common with a similar pressurising connection in the splice box area.

hoses) which previously could be manufactured only using a batch process in autoclaves.

7. Problems associated with die-face cooling are eliminated because the cooling medium is remote from the extruder die head.

8. In the case of XLP power cables the elimination of steam gives a 'dry cure' to the cable, considerably reducing the number and size of microvoids.
9. The higher usable temperature permits higher line speeds.
10. Because of the buoyancy of the salt the tube is straight, a catenary not being necessary (specific gravity of liquid salt is approximately 1·9).

### 9.2 Lead Extrusion

Another important continuous operation introduced by the cable industry has been the continuous extrusion of lead by a lead extruder[11,12] rather than by a lead press; this has been well established and increasing in usage since the mid-1950s.

## 10 THE FUTURE

This chapter has been restricted to polymeric cables in general usage. For future applications as modern industrial processes become more continuous, automatic and sophisticated, there will be parallel developments within the cable industry to adapt existing designs or indeed develop completely new cables, should these be needed for specific purposes.

More generally there appears to be a growing demand for higher operating temperatures with similar reliability to existing cables. Greater flame resistance, less toxicity and less smoke emission on burning is also an ongoing development requirement worldwide. The cable industry already uses most polymers and thermoplastic materials and of necessity it will continue to search for improved polymeric materials and will not hesitate to use them if they are suitable.

This equally applies to machinery and especially to vulcanisation and braiding. The latter will now be discussed.

## 11 MACHINERY

### 11.1 Hose and Cable Braiding Equipment

Three completely new braiders have been designed and announced recently. They have successfully completed field factory proving trials and are now commercially available for use with textiles, including

Kevlar Aramid.[7] These excellent machines are: (1) the Rockwell RB2 Horizontal Rotary Braider, (2) the Karg High Speed Vertical Rotary Braider and (3) the Mayer Rothkopf MR11 Horizontal Braidmatic Braider.[13]

All three machines are also suitable for braiding with conventional brass-coated steels and stainless steel wires as used in the manufacture of high-pressure hydraulic hoses. The Rockwell and Karg machines are currently undergoing field factory testing, while the Mayer Rothkopf machine is immediately available. This latter machine also has the unique capability of being able to braid hoses of diameters larger than 0·75 in without having to follow standard procedures of constructions manufactured on 36- and 48-spindle braiding machines. Indeed, it is highly likely that for the vast majority of sizes of hydraulic hoses of the wire-reinforced varieties, only 16-, 20- and 24-spindle braiders will be necessary, which will be a major step forward for rationalisation within hose-manufacturing plants. The operating principles of these machines will now be detailed, together with their main advantages. It will, of course, be necessary for individual hose and cable manufacturers to ascertain, determine and choose for themselves which machine is most suitable for their own particular requirements, bearing in mind the types of hoses required and their own already existing equipment.

### *11.1.1 The Rockwell RB2 Horizontal Rotary Braider*

This braider is rotary in principle, and also employs mechanical means to displace outer carrier strands to effect interlacing. The design permits a reduction of speed-restrictive inertial forces and allows for increased bobbin capacities. A power-former is positioned at each braiding point, which produces a uniform 'locked-in' braid. Being rotary in principle, the maximum and minimum tension variations, which can cause problems to the hose designer, brought about by the maypole or figure-8 movement of the carriers during braiding, are thus eliminated. It should also be possible to use reduced braiding tensions as there is now no low-tension point to require consideration.

### *11.1.2 The Karg High Speed Vertical Rotary Braider*

Being rotary in principle, this machine has similar advantages to those discussed in the preceding paragraph, when compared with machines using the figure-8 movement configuration, but being vertical and with an optional design capability of speed synchronisation with a second vertical machine, it is thus possible to use as a twin unit for two-braid hoses, or to use each deck separately and simultaneously for. one-braid hoses, thus giving maximum deck utilisation to the hose production departments. This is not possible with twin-deck horizontal units and is a big production advantage for single-wire braiding.

*11.1.3 The Mayer Rothkopf MR11 Horizontal Braidmatic Braider*[13] (Fig. 2)

A design feature of this machine is the elimination of angular acceleration because the yarn/wire ribbons point directly at the hose coming from the carrier payoff roller at any position during its inward or outward cycle. Planetary gears compensate for the angular movement, preventing the payoff roller from swinging from one side to the other. Thus the ribbons are positively guided and the chance of cross-overs reduced. There is practically no movement of the hose at the braiding point during braiding because of the equal tension of opposite carriers and therefore no disturbance caused by the swinging of the ribbons.

Another design feature is that during the transfer and at any other

FIG. 2. MR11 Horizontal Braidmatic Braider. (Courtesy: Mayer Rothkopf Industries)

point of the braiding cycle, the shuttle is guided positively. This latter item features two semicircular cam tracks, while the rotor includes two back and two front cam followers, which run on the cam tracks and hold in the shuttles in their respective positions.

A hidden design feature is the use of roller bearings for all moving surfaces, and these, coupled with the shuttle movement, drastically and noticeably reduce energy consumption in usage.

Being a braider of maypole configurational movement, there are of course highs and lows in the payoff tensions of the ribbons, but these have been considerably reduced and narrowed by a novel tension control incorporating pulley wheels. The configuration and principle adapted permits the 'slack' between the inner and outer cycle to be stored by turning the compensator plate, which itself is under tension. As the plate turns, the tension increases or decreases gradually.

**11.2 MR11 Double Let-off Braiding System** (Figs. 3 and 4)
A Mayer Rothkopf braiding machine fitted with what is in effect an additional yarn or wire bobbin riding 'piggy-back' allows the use of more ends in the ribbon, to give the extra coverage needed on the circumference of larger diameter hoses. This is usually achieved by the use of additional carriers or spindles in the braiding machines, making it necessary to use 36–48-spindle machines. By using a 24-spindle MR11 machine fitted with the double let-off system, wire-braided hoses have been manufactured and satisfactorily tested by several well-known hose

FIG. 3. MR11 double let-off wire machine. (Courtesy: Mayer Rothkopf Industries)

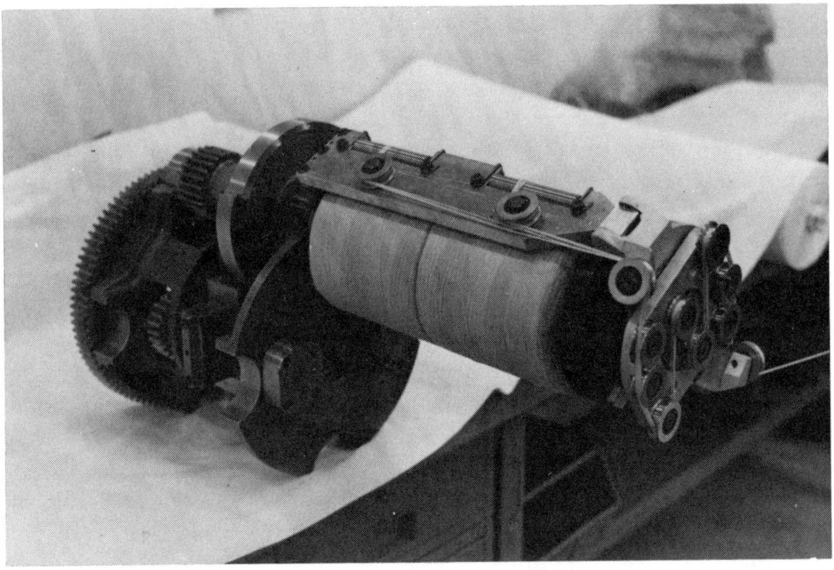

FIG. 4 MR11 double let-off textile machine. (Courtesy: Mayer Rothkopf Industries)

manufacturers, in sizes which would normally require many more carriers. The impulse test results obtained for such hoses were also higher in number, which is not really surprising when one considers that there are fewer wire cross-overs and hence fewer rubbing points on such a hose, when compared with the normal procedures requiring more carriers, size for size.

The three machines discussed, in their various spheres and ways, will all play a major role in future hose and cable production units, and all have major technological breakthroughs in their designs when compared with the existing slower and conventional braiders which have been in use for many years.

## REFERENCES

1. Duprene, Neoprene. Registered Trade Names, Du Pont de Nemours, Wilmington, Delaware, USA.
2. BENNETT, H. R., MASCHIO, G., MCCONNELL, J. and PURVIS, J. (Pirelli General Cables) (1975) The contribution of elastomeric insulated cables to

the solution of energy problems. Presented at 3rd PRI National Conference, London.
3. MORELLO, A., TARALLI, C. and MCCONNELL, J. (1975). Power transmission by way of submarine cables. Offshore Technology Conference, Houston, Paper No. OTC 2257.
4. EVANS, C. W. (1979) *Hose Technology*, 2nd Edn, Applied Science Publishers London.
5. EVANS, C. W. (1978) *Powdered and Particulate Rubber Technology*, Applied Science Publishers, London.
6. EVANS, C. W. (1980) Hoses for the offshore oil industry. Presented at Norges Gummi Tekniske Forening Conference, Stavanger.
7. Kevlar Aramid. Registered Trade Name, Du Pont de Nemours, Wilmington, Delaware, USA.
8. COWAN, J. and MANLEY, T. R. (1976) The effects of inorganic borates on the burning of PVC. *Br. Polym. J.*, June.
9. NYE, H. F. (BICC) (1975) Service requirements for modern cables. Presented at 3rd PRI National Conference, London.
10. *Continuous Vulcanising Systems for Rubber and XLP Cables*, General Engineering Co. Ltd, Radcliffe, Manchester.
11. Continuous Lead Extruder, H. Folke-Sandelin, Stockholm, Sweden.
12. Continuous Lead Extruder, Pirelli, Milan, Italy.
13. Braidmatic. Registered Trade Name, Mayer Rothkopf Inc., Orangeburg, South Carolina, USA.

*Chapter 2*

# POLYMERS AND FOOTWEAR MATERIALS

A. R. CARTER and R. E. WHITTAKER

*Shoe and Allied Trades Research Association,*
*Kettering, Northants, UK*

*SUMMARY*

*The footwear industry today relies very much on the polymer industry for shoe solings, uppers, adhesives and components. In the UK only about 5% of shoe soles are currently made of leather. This chapter considers the requirements and performance of the alternative soling compounds of which rubber in its various forms still occupies a large volume. The use of two-compound moulding is discussed with particular reference to the relatively new applications of rubber in this area. The development of PVC- and polyurethane-coated fabric upper materials is dealt with together with the poromerics. Although the use of these types of materials as shoe uppers peaked in the mid-1970s, they still remain a significant part of the market and recent developments may result in further penetration.*

## 1 INTRODUCTION

Over the last 50 years polymers have had a considerable effect on the footwear manufacturing industry. Most of the major process developments that have occurred in the industry during this period have been due to polymer developments. Prior to 1930 the whole of footwear was made from leather and the soles were attached to the upper by sewing, which made the industry very highly labour-intensive.

The first moves took place before the last war with the early development of adhesives and sheet rubber solings. This combination led to a dramatic change in production in that for the first time sewing was not required to join the upper to the sole and the stuck-on process was developed. The early soling developments were based on natural and synthetic rubber with a high content of resin. The latter type of material became known in the footwear industry as resin rubber and is still widely used today as a sheet soling material in men's and women's footwear.

The escalation of direct vulcanising (DV moulding) in the 1950s led to a further development in improved processing of footwear. Moulding as a means of joining upper to sole was introduced into the industry. This move was further accentuated in the 1960s with the development of PVC and plastics injection moulding techniques.

The development of polyurethane in the 1970s brought a lightweight, relatively highly durable fashionable sole to the industry and also introduced the process of reaction moulding. The combination of these developments in adhesives and solings has meant that only 5–6% of UK-produced shoes had leather soles in 1980 and only about 4–5% of the soles were stitched to uppers, this being done mainly by the welted shoe process.

The development of polymeric-based materials to replace leather as a satisfactory upper material for shoes has been a more difficult and slower development. Whereas 95% of the footwear produced in the UK in 1979 had polymeric-based soling materials, only 17·8% had synthetic upper materials, as shown in Table 1. The proportion of shoes produced with synthetic uppers reached a peak of 27·8% in 1974. The main reason for this low proportion of synthetic upper materials is the inability to compete with leather in moisture absorption and permeability properties. For this reason, most of the use of synthetic materials is confined to women's footwear where, because of the generally more open construction of the shoe, permeability of the shoe upper is not too important a property.

## 2 SHOE UPPER MATERIALS

As well as good comfort and hygienic properties, leather is also very tough, reasonably flexible and has good durability. Despite these good properties, however, it does suffer from a number of disadvantages. It cannot for example be obtained in uniform rolls and this has held back

TABLE 1

UK FOOTWEAR PRODUCTION, BY SYNTHETIC UPPERS (MILLION PAIRS)

| | *1968* | *1969* | *1970* | *1971* | *1972* | *1973* | *1974* | *1975* | *1976* | *1977* | *1978* | *1979* | *1980* | *1981*[a] |
|---|---|---|---|---|---|---|---|---|---|---|---|---|---|---|
| Poromerics: | | | | | | | | | | | | | | |
| men's | 0·3 | 0·9 | 1·5 | 1·9 | 2·8 | 4·1 | 4·7 | 2·7 | 2·9 | 2·8 | 2·6 | 1·9 | 1·6 | 1·4 |
| women's | 0·3 | 1·1 | 2·6 | 4·4 | 3·2 | 6·5 | 2·6 | 0·8 | 1·1 | 1·1 | 0·7 | 0·2 | 0·2 | 0·2 |
| children's | 0·1 | 0·5 | 0·9 | 2·2 | 1·5 | 1·7 | 2·5 | 1·0 | 0·3 | 0·3 | 0·2 | 0·2 | 0·2 | 0·2 |
| Total | 0·7 | 2·5 | 5·0 | 8·5 | 7·5 | 12·3 | 9·8 | 4·5 | 4·3 | 4·2 | 3·5 | 2·3 | 2·0 | 1·8 |
| Coated fabrics: | | | | | | | | | | | | | | |
| men's | 0·4 | 0·3 | 0·4 | 0·5 | 0·4 | 1·0 | 1·6 | 2·9 | 2·4 | 3·2 | 2·6 | 1·2 | 0·8 | 0·6 |
| women's | 21·2 | 21·4 | 26·7 | 28·1 | 25·4 | 31·4 | 30·7 | 28·1 | 23·5 | 22·0 | 22·0 | 19·7 | 20·1 | 20·4 |
| children's | 6·0 | 6·0 | 6·2 | 6·6 | 6·5 | 6·1 | 6·1 | 5·6 | 3·9 | 4·6 | 3·4 | 3·2 | 2·4 | 2·3 |
| Total | 27·6 | 27·7 | 33·3 | 35·2 | 32·3 | 38·5 | 38·4 | 36·6 | 29·8 | 29·8 | 28·0 | 24·4 | 23·3 | 23·3 |
| All synthetics: | | | | | | | | | | | | | | |
| men's | 0·7 | 1·2 | 1·9 | 2·4 | 3·2 | 5·2 | 6·3 | 5·6 | 5·4 | 6·0 | 5·2 | 3·1 | 2·4 | 2·0 |
| women's | 21·5 | 22·5 | 29·3 | 32·5 | 28·6 | 37·9 | 33·3 | 28·9 | 24·6 | 23·1 | 22·7 | 19·9 | 20·3 | 20·6 |
| children's | 6·1 | 6·5 | 7·1 | 8·8 | 8·0 | 7·8 | 8·6 | 6·6 | 4·3 | 4·9 | 3·6 | 3·4 | 2·6 | 2·5 |
| Total | 28·3 | 30·2 | 38·3 | 43·7 | 39·8 | 50·9 | 48·2 | 41·1 | 34·3 | 34·0 | 31·5 | 26·4 | 25·3 | 25·1 |
| Total footwear production | 200·8 | 194·1 | 190·5 | 194·2 | 184·4 | 189·2 | 173·2 | 164·0 | 157·2 | 159·6 | 154·7 | 149·6 | 136·0 | 128·7 |
| % of footwear with synthetic uppers | 14·1 | 15·6 | 20·1 | 22·5 | 21·5 | 26·8 | 27·8 | 26·2 | 21·6 | 21·3 | 20·3 | 17·8 | 18·6 | 19·5 |

[a] Provisional figures.

Source: Business Monitor PQ 450 and SATRA estimates, December 1981.

Note: Analysis of poromeric and coated fabric data is estimated from 1978 onwards.

automation in shoe upper construction. Also, serious faults within leather are caused by insects such as warble fly whose larvae live in the animal skin. Fungal infections, e.g. ringworm, can also cause damage, as do scratches from barbed wire fences. Variations in leather quality are also caused by the techniques of husbandry, environment and age of the animal. These faults, however, have to be masked by the tanner as best as possible, but it is not always done satisfactorily. A certain amount of wastage must result from this treatment and this is the main reason why footwear and clothing manufacturers have increasingly looked to the possibility of using synthetic materials which can be obtained in rolls and also have fairly uniform physical properties.

Patents relating to methods and techniques of preparing various synthetic leather compositions have been granted to inventors as far back as 1850; even before that time, a number of attempts were made to devise a simple and rapid technique of producing synthetic leather compositions. In most of the early art, pyroxylin was used to coat or impregnate various types of fibrous base materials to prepare leather substitutes. As time went on, pyroxylin/oil/pigment compositions were widely used as coating or impregnating compositions for various woven or non-woven fibrous materials.

The outstanding moisture absorption and permeability properties of leather are due to the molecular and physical structure of its protein precursor, collagen. Collagen is the main constituent of all structural tissue and is converted into leather by treatment with such tanning agents as chromium salts, vegetable tannins and aldehydes. The most hydrophilic man-made polymer yet manufactured, however, has not managed to absorb more water than leather itself. This fact led some materials manufacturers to the idea of using collagen or leather fibres as a possible raw material for the manufacture of artificial leather. This class of materials, known as collagenous poromerics, were developed around 1970 but were not taken up by the industry mainly because of price.

## 3 USAGE OF SYNTHETIC UPPER MATERIALS

The main types and usage of synthetic upper materials are shown in Table 1. PVC-coated fabrics started to be used by the footwear industry around 1960 and this has increased each year up to the mid-1970s. Polyurethane-coated fabrics with their improved handle properties started to develop in the late 1960s and now represent the major type of coated fabric material being used in the industry.

The attempt to challenge leather as the predominant upper material began in 1964 with the marketing of 'Corfam' poromerics manufactured by E. I. Du Pont de Nemours Inc. in the USA. The rapid development of permeable synthetic upper materials under the generic name of 'poromerics' reached its peak around 1969–70 and most of the major chemical companies undertook various developments at certain times. Despite the great expectations ten years ago, developments over the intervening period have shown that the challenge to leather has been very satisfactorily resisted. Over the last seven years the percentage of poromerics used has gradually decreased and really only two, 'Clarino' manufactured in Japan and 'Porvair' manufactured in the UK, are used in any large quantities today.

## 4 PVC-COATED FABRICS

The first group of man-made upper materials to be used by the footwear industry were impermeable PVC-coated fabrics. There are very many types of coated fabrics,[1] but the PVC materials used most frequently in the footwear industry are either solid or expanded, supported upon a woven, knitted or impregnated non-woven fabric base. The fibres most widely used for the base fabric are cotton and nylon.

The plastic coating on the fabric provides the attractive finish, which can be coloured, printed and embossed to provide many varied effects, but is usually required to be leather-like. It provides the good wearing properties of the product, and it must be able to withstand a considerable amount of abrasion and flexing and retain a good appearance. The nature of the polymer coating affects the handle of the product, in that the coating may be stiff and harsh, or soft and supple, according to requirements. The fabric provides most of the tensile strength and almost the whole of the tear strength of the product. The fabric also considerably modifies the stretch properties of the coating and makes a major contribution to the handle of the material. Interaction effects between plastic and fabric are pronounced, e.g. the tear strength of the composite may be less than that of the strongest component, but the scientific study of the properties of flexible composites has been limited.

The major problem with PVC-coated fabric materials is that in most cases they are impermeable and thus prevent sweat escaping. The sweat materials accumulate at the junction between the coating and the fabric and can cause delamination. Microporous PVC coatings have been developed which are permeable to water vapour, but they suffer from poor surface abrasion properties and have generally been limited to use as footwear lining materials.

Because PVC flows under high-frequency heating, it was possible to develop specially the process of flow moulding in 1970. In this process a silicone rubber mould is made by reproducing the contours of a finished leather shoe upper. The PVC-coated fabric is then placed in contact with the mould under a high-frequency press and an upper design is embossed on the plain-coated fabric material. The process has been used widely in the production of cheaper ladies' shoes. Some men's and children's shoes have also been produced.

## 5 POLYURETHANE-COATED FABRICS

Polyurethane-coated fabrics (commonly referred to as PUCFs) have been widely used in the footwear industry over the last ten years.[2] Their popularity arises from the leather-like handle, appearance, flexibility, availability in a wide range of aesthetically appealing finishes, and reasonable price. Polyurethane-coated knit back materials were the first type of PUCF to be used in footwear. They fitted well into the demand for a tight-fitting stretch leg boot in 1969–70 and were widely used with some problems. The fashion at the time made it necessary to stretch the stitched fronts of the boots by about 40% to put them on the foot and this caused considerable returns from the customer due to splitting and tearing failure.

The main growth in PUCF materials has come with the development and use of transfer-coated materials having a thin solid polyurethane coating put on to raised cotton or polyester–cotton woven fabrics, and these materials constitute the bulk of PUCFs used in the last ten years. More recently, however, coagulated and expanded PUCFs offering improved properties have been developed and used.

The structure of a conventional PUCF material is shown in Fig. 1. There are generally three main components:

- (i) a thin surface (grain) film of polyurethane polymer about 0·05 mm (0·002 in) thick;
- (ii) a polyurethane adhesive layer which attaches the polyurethane (PU) film to the substrate fabric;

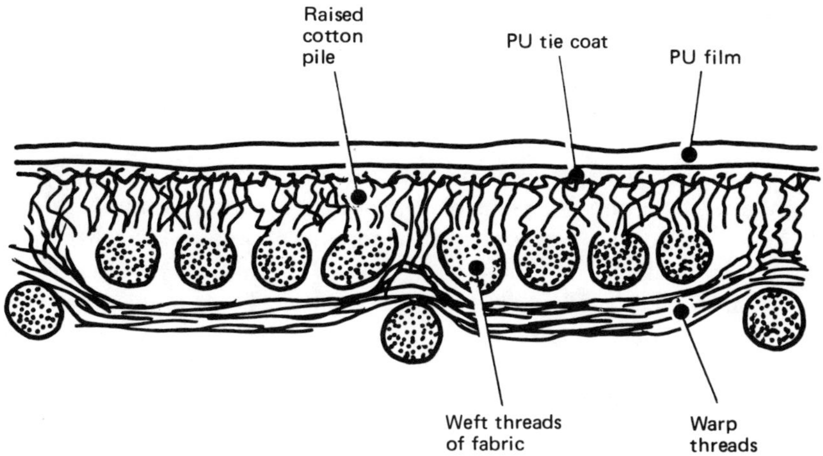


FIG. 1. Structure of conventional polyurethane-coated fabric.

(iii) a raised fabric which may be cotton or a polyester–cotton blend; this is the substrate which provides the bulk of the mechanical strength of the final PUCF.

The unique attractive look and handle of a standard PUCF relies on the flexible link between the PU film and the weave of the fabric. This is achieved by the raised pile fibres, to which the film is attached, which allow the PU film to 'float' above the fabric weave and to behave almost independently of it. It is this structural feature which produces the very attractive leather-like break in PUCFs. This is further enhanced by the various fine finishes and gloss coats applied to the PU film at the final stage in manufacture.

The original method of making PU-coated fabrics was to coat the solution of PU compound directly on to the substrate fabric, and such materials were used in the manufacture of protective clothing. The aesthetic properties of these early PUCFs were not good enough for fashion garments or for footwear. Experiments in the PU coating of raised cotton fabrics demonstrated the leather-like appearance that such materials possess and their attractiveness and good handle was soon recognised as important for footwear uppers, garments and upholstery.

The key to the properties of this particular type of PUCF is the quality and uniformity of the cotton pile. In the early materials, adequate quality of pile was not consistently obtained. The value of the two other developments, coagulated and foam coatings, is that the unique properties of these PUCFs do not rely upon the presence of a raised cotton pile.[3] In coagulated coatings the pile is replaced by the special nature of the coagulated PU and in foam coatings it is replaced by the soft, flexible PU foam, both of which give similar desirable properties in the final product.

The coagulation process relies on the impregnation of the fabric component, commonly a plain or raised woven cotton fabric, with the PU dissolved in the powerful and toxic solvent known as DMF (dimethyl formamide). The fabric to be coated is passed between rollers immersed in a bath of low-viscosity PU solution. This pre-impregnation stage ensures good bonding of the main coat, which may be applied by another roller stage or by knife coating over a steel roller. The coated/impregnated fabric is then passed through a number of baths of water or brine containing a proportion of DMF to assist the control of pore size. The DMF in the PU solution is leached out of the coating into the water by osmotic forces. This makes the polyurethane coagulate as a porous layer on and in the fabric. The material is then dried and the surface flattened prior to spraying, lacquering and embossing. Foam PUCF involves the transfer coating of the thin solid PU film using a PU foam interlayer on to plain woven or non-woven fabrics.

It is the coagulated PUCFs which have attracted the most interest recently because of their excellent soft handle and their extremely leather-like appearance in the shoe. In many respects these materials are simplified poromeric upper materials and some fall within the accepted definition of a poromeric.

## 6 POROMERICS

### 6.1 Fibrous Poromerics

The development and marketing of a permeable synthetic material under the generic name of a 'poromeric' has not been a great commercial success story. Most major chemical companies (e.g. ICI, Goodrich, Du Pont, AKZO) have at some time spent considerable sums in research and development on new materials and few have received much return on these investments. The main reason why these materials have not been able to compete with leather is due to a number of factors, particularly price and comfort properties.

The manufacture of a completely synthetic material which had moisture absorption and permeability properties similar to those of the natural leather was started in the late 1930s by E. I. Du Pont de Nemours Inc. in the USA. They initially used the term 'poromeric' to describe the material and defined it as 'a microporous permeable, coriaceous sheet material comprising polyurethane reinforced with polyester'. Commercial production of the Du Pont product, under the trade name of 'Corfam', was not started until 1964, but as early as 1942 a US patent[4] was taken out to describe a shoe upper material composed of flexible fabric coated with a synthetic linear polyamide.

A number of poromeric upper materials have come on to the market in recent years. In view of the development of these materials, the Shoe and Allied Trades Research Association widened the first definition given by Du Pont and defined a poromeric material used in footwear as 'a man-made shoe upper material which is generally similar in nature and appearance to leather and in particular has a comparable permeability to water vapour'.

Corfam in the initial stages contained a woven interlayer (Fig. 2) as well as a main fibrous substrate layer and hence was rather stiff to wear and lacked the permeability/absorption characteristics necessary for men's shoes. Although softer grades were introduced, production at the USA plant stopped in 1971. The plant and expertise was sold to Poland and a similar material is being marketed under the name of 'Polcorfam'.

In general, poromerics[5–7] had at least two discrete layers, a dense microporous linear polyurethane layer making up 20–35% of the total thickness and a polyurethane-impregnated fibrous layer of 65–80% of the thickness. In the case of Corfam, five layers could probably be identified:

(a) *Finish (or top) layer*. This is a very thin film (about 0·001 mm thick) applied to the otherwise complete poromeric. These finishes are commonly acrylic polymers but may sometimes be transparent polyurethane.

(b) *Surface skin*. This is usually an integral part of the microporous layer, being formed as a result of fusion of the outermost cells. In some instances this surface is formed during a film release process and in others as part of the grain embossing process. The thickness (usually about 0·005 mm) and properties of this skin are very important because they affect the resistance to water penetration and the water vapour permeability of the poromeric, also resistance to scuffing and snagging damage and, in combination with the microporous coating, determine the type of 'break' (wrinkle formation) that the material will form when folded.

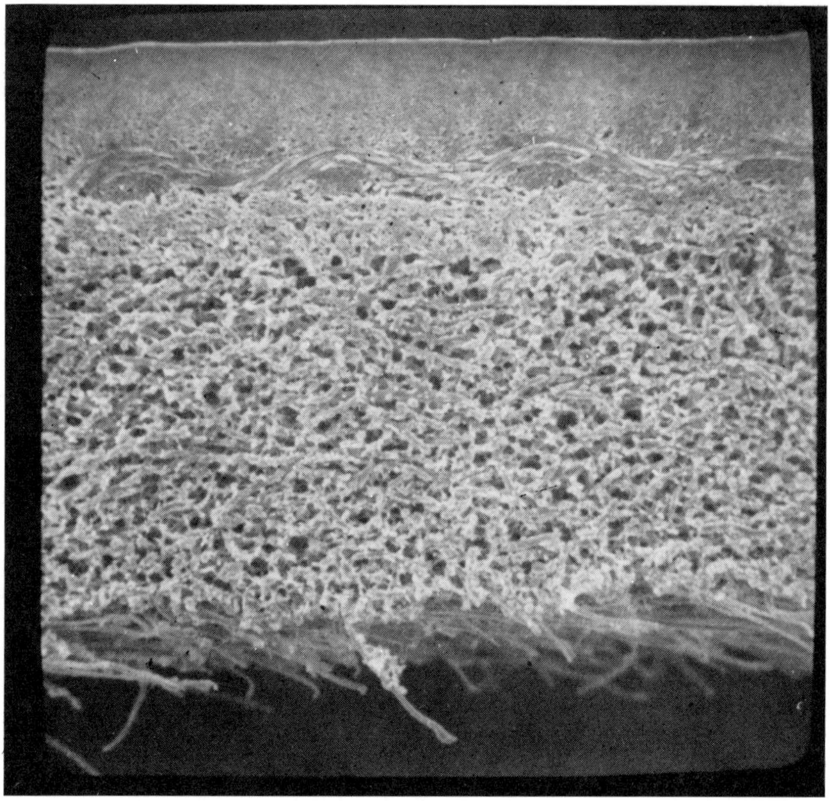

FIG. 2. Stereoscan of poromeric 'Corfam'.

(c) *Microporous layer*. This layer is generally a permeable foam of small cell dimensions, although several different types of cellular structures are used in different materials.

(d) *The interlayer*. This is commonly a woven fabric which separates the microporous layer from the non-woven substrate in some materials. This component serves several purposes, e.g. it increases tensile and tear strength (but reduces breaking extension), controls stretch and creep, masks surface unevenness effects when the material is strained, and places the neutral plane of the material close to the surface skin. The interlayer fabric in Corfam was polyester but various fibres are used in different poromerics. Although such layers provide the benefits listed above, their use creates the problem of delamination which has caused some difficulties in shoe manufacture.

(e) *The substrate.* This can be a non-woven, woven or knitted fabric, all of which are impregnated to various extents with a polyurethane elastomeric binder. However, in one instance, discussed below, the substrate is cellular polyurethane only. In general, the impregnated substrates tend to behave like the fabric reinforcement which they contain. Thus, substrates with a non-woven matrix will have a high breaking extension and tend to creep, whereas those substrates with woven fabrics will have lower breaking extensions and will tend not to creep; knitted substrates come somewhere between the two.

One of the most successful poromerics has been 'Clarino', which was developed in the late 1960s by Kuraray Corporation in Japan. This has always been manufactured without an interlayer and consists of a microporous polyurethane foam and fibrous base. Generally a large amount of foam elastomer was included in the substrate and this contributed to the good handle properties. A new grade of Clarino, under the name of Clarino F, was developed in 1975 and has been increasing in production. The main difference is that Clarino F uses a special process to produce very fine microfibres in order to give it a very soft leather-like feel.

### 6.2 Homogeneous Poromerics

The structure of 'Porvair', manufactured at King's Lynn in the UK since 1967 by the company with the same name, is different from all the other poromerics in that it does not contain any fibre. A stereoscan photograph is shown in Fig. 3; in basic form it consists of just a polyurethane skin and microporous polyurethane foam.[5–7] The foam usually has a gradation in pore size through its thickness. The modulus of the material is also much lower than with the fibrous poromerics and it has good surface abrasion properties. The material is also different from other poromerics in that it displays viscoelastic behaviour and acts as a polymer with time- and temperature-dependent physical properties. The breaking extension of the material is over 400%, which may be compared with values of under 100% for most of the fibre-based materials.

One of the disadvantages of Porvair, which follows from the absence of an interlayer or fibrous base, is its high stretch and liability to tear if handled like leather. Therefore care has to be taken when making leather and Porvair shoes in a factory at the same time. For this reason a

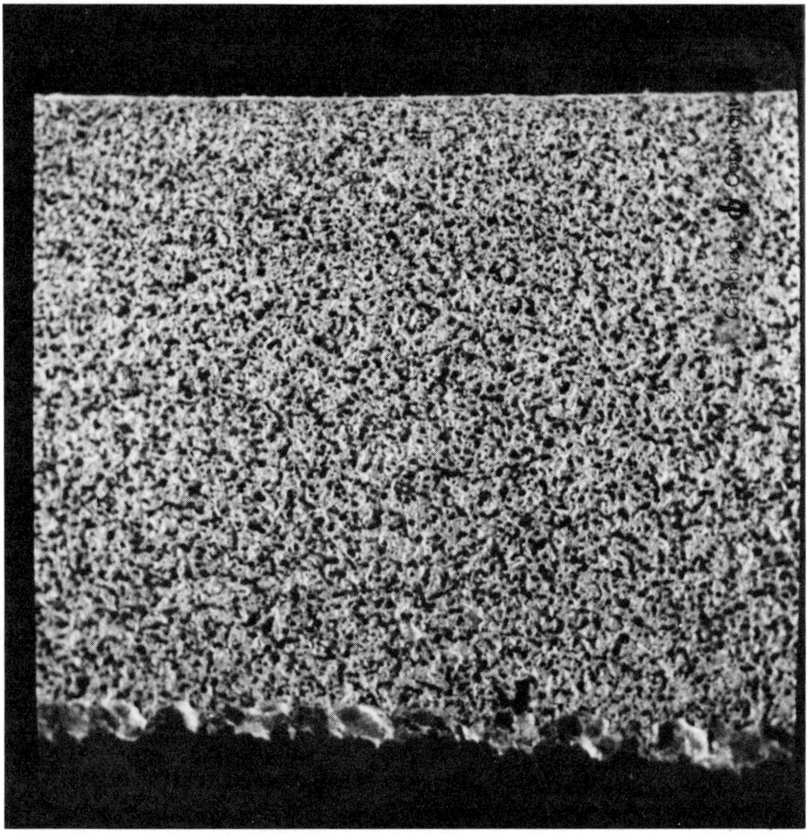

FIG. 3. Stereoscan of poromeric 'Porvair'.

modification under the name of Vantel is now being produced which contains a non-woven backing material laminated to the back of conventional Porvair to prevent too much stretch in shoemaking.

The advent of a homogeneous poromeric to the footwear industry has introduced various plastic fabrication techniques into shoe production such as high-frequency welding, embossing, edge burnishing, vacuum and pressure forming.

### 6.3 Strength of Poromeric Materials

One of the most surprising features in the use of foam polyurethane as an upper material (and cellular polyurethane in soling materials) is its high strength when compared with vulcanised rubbers.[8,9] The stress–strain curves for foam and solid polyurethane and vulcanised natural rubber are shown in Fig. 4. It would appear that the initial modulus of the urethane foam is higher although the actual tensile strength is lower than a solid NR vulcanisate. The modulus of the solid polyurethane is extremely high when compared with the corresponding foam and its tensile strength is considerably in excess of that found in a natural rubber

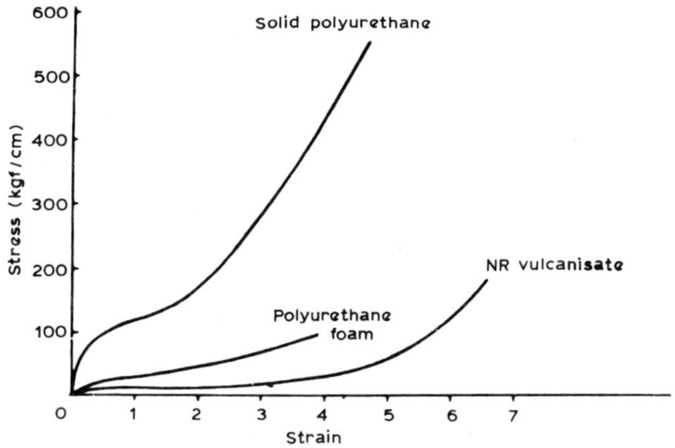


FIG. 4. Stress–strain curves for NR vulcanisate, foam and solid polyurethane materials.

vulcanisate. The other surprising feature is shown in Fig. 5, where energy input to break is plotted against temperature for foam and solid polyurethane and filled and unfilled SBR vulcanisate. The very high strength of the linear polyurethanes used in poromerics is retained over a wide temperature range and only reduces at temperatures above 170 °C where a structural change occurs in the polyurethane; this is usually outside of the temperatures reached in shoe factory processing.

The high strength and abrasion resistance of linear polyurethanes is due to their chemical structure. The materials are composed of hard urethane segments of very minute size in the polyester or polyether rubber matrix. These hard urethane segments act as minute filler particles in the rubber matrix to produce a very effective 'self-reinforced' elastomer. Some problems with poromerics have occurred due to hydrolytic failure of the polyurethane foam[7–10] in wear, but progressive improvements by the suppliers have minimised this type of failure in recent years.

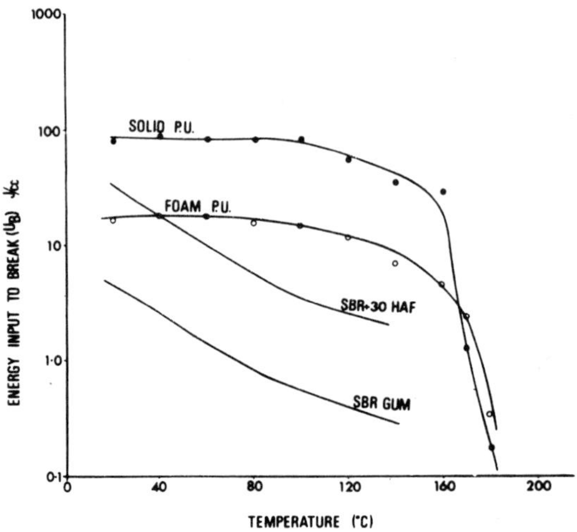


FIG. 5. Variation of energy input to break with temperature for foam and solid polyurethane, SBR unfilled and filled with 30 pphr carbon black.

## 7 POLYMERS IN SHOE COMPONENTS

Although the main uses of polymers have been in the soling and upper materials in footwear, a number of rubber and plastics materials have been used as shoe components. Polystyrene, for example, is the main polymer used in the manufacture of shoe heels. In the last few years fashion trends and changes have had a considerable effect on the industry and polymers have greatly assisted designers in meeting the challenge. During the platform/clog fashion boom from 1972 to 1974, PVC, rigid and flexible polyurethane, ABS, polystyrene and other polymers were used. Polystyrene and Surlyn have been used as toe puffs and stiffener materials and experimental insole materials have been made up.

## 8 SHOE SOLING MATERIALS

Traditionally footwear solings were basically functional. They were used to protect the bottom of the foot from sharp objects on the ground and thus allowed greater mobility. In early footwear the upper part of the shoe was also functional in that its main objective was to hold the sole in contact with the foot, and the same principle is still employed with the simple beach sandals currently available.

Today, of course, the soling still maintains the functional role. However, in the current competitive and demanding market the soling is additionally expected to have a degree of fashion appeal and this trend towards improved aesthetics has been very apparent during the past ten years.

**8.1 Use of Rubber in Footwear Solings**

In 1979 about 47% of solings were manufactured from rubber. It is estimated that 20 000–25 000 tons of rubber compound is currently used in footwear. However, although the bulk of this is used in solings, this includes separate heels and some may also be used by the repair industry.

During the decade the proportion of rubber soles has decreased from about 54% in 1971. The main reason for this trend has been the substitution of rubber by plastic solings, particularly the reaction-moulded polyurethane and thermoplastic rubber types. In fact, because of the decline in UK-produced footwear during this period, the shortfall on rubber-based solings during the decade is more than 30 million pairs. This situation is shown in Fig. 6, which relates rubber solings to total solings over the period 1971–81. The direct-moulded rubber production shown in Fig. 6 involves almost entirely a compression moulding technique using either a solid compound, or one containing a blowing agent for slipper production. The vulcanised rubber sheeting and units occupy the larger pairage and of these the rubber sheet soling is by far the most important with more than 33 million pairs in 1981.

Currently all sheet solings occupy 34% of total solings and the proportion of rubber in this sheet soling market is about 77%. The UK usage of sheet solings increased from 31% in 1979 to 34% in 1981, and although they are traditional solings, the reason for their present popularity is due to the current fashion demand for an elegant shoe with a thinner sole and practical considerations. In the current footwear manufacturing environment, short production runs and the necessity of rapid style changes predominate. Sheet solings are ideal in this situation as there is no need to invest in moulds and no reliance on mould manufacturers. The tooling is inexpensive as the basic requirement is a knife to cut the soling to the desired shape. This of course enables a very quick response to fashion changes. Also, by bonding different sheet materials together, a very attractive and light soling may be obtained.

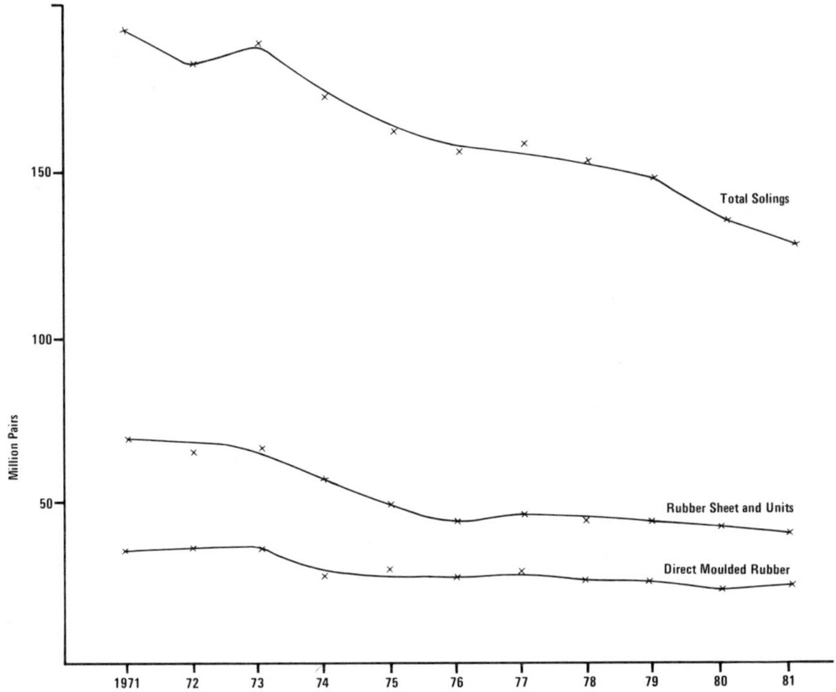


FIG. 6. Rubber in relation to total solings, 1971–81.

## 9 PERFORMANCE OF INDIVIDUAL SOLINGS

### 9.1 Direct-Moulded Solid Rubber

This is the oldest method of moulding a sole directly on to the upper part of the shoe and dates back to the 1930s. The upper is mounted on a metal foot which is the male part of the mould. A suitable adhesive is applied to the bonding area and a prewarmed unvulcanised rubber blank is placed in the heated metal sole mould which has a cavity shaped in the form of a shoe sole. The mould is closed, pressure is applied and the total vulcanising time is about 8 min at a temperature of 150–160 °C. The actual moulding conditions are dependent on the type of compound and the dimensions of the sole. The compounds used are quite complex and often contain more than twenty components in the mix. The base polymers used for normal production are SBR or NR, but when oil resistance is a requirement, then nitrile rubber and Neoprene rubber are used to varying extents.

The solings produced by this method are mostly used on industrial and safety boots. The type of sole design used in this footwear incorporates deep cleats to ensure adequate traction, but this design feature can result in high localised surface strains during normal walking. To avoid flex cracking problems in service, the compound therefore has to be of quite a high quality.

In practice very few problems occur with this type of soling. In the past, adhesion failure between the upper and the sole sometimes used to occur, but today this can normally be avoided by controlling the content and type of grease in the leather upper materials.

### 9.2 Direct-Moulded Sponge Rubber

This moulding process again employs a compression technique and is quite similar to that used with the solid rubber, the essential difference being in the compound and the moulding conditions employed. This soling is used mainly in the production of household slippers. Usually the upper is based on a fabric which is less sensitive to heat than leather. It is therefore possible to use a hot moulding foot at a temperature of about 130 °C as well as the heated sole mould. As the cellular soling compound has relatively poor wearing properties, a thin solid semi-cured wearing surface is initially placed in the sole mould. The preweighed blank of cellular compound is positioned on top of the solid skin and the mould is closed. The sole mould temperature is normally about 160 °C and the compound, which is totally surrounded by hot metal, expands as the blowing agent reaches its activation temperature. No external pressure is applied other than that necessary to keep the mould closed, as the pressure exerted by the blowing of the compound is sufficient to make the bond to the upper and to form the shape of the sole.

Wear problems with slippers are minimal. The solid wearing surface gives adequate performance under normal service conditions and even when this skin is worn through, the cellular rubber still has some residual wear.

### 9.3 Sheet Solings

As mentioned previously this category of solings is very important, especially in today's trading environment, and the full range of common sheet solings is shown in Table 2. The approximate density of each soling is also given, and this is a particularly important basic property of the soling especially when in the cellular form, where the lightness of the footwear may be a selling feature. The main properties of these solings are summarised in Table 3 and discussed in more detail in the following subsections.

TABLE 2
SHEET SOLINGS IN UK FOOTWEAR PRODUCTION IN 1981 (ESTIMATED)

| *Type of soling* | *Million pairs* | *Typical density ($g/cm^3$)* |
|---|---|---|
| Resin rubber | 25·8 | 1·3 |
| Microcellular rubber | 4·7 | 0·6 |
| Gristle rubber | 1·0 | 1·2 |
| Crepe rubber | 2·1 | 0·9 |
| Leather | 7·1 | 1·0 |
| EVA (ethylene vinyl acetate) | 2·7 | 0·3 |

### *9.3.1 Leather*

This is of course the traditional shoe soling but today is used on only about 5% of the shoes produced in the UK. The durability of leather is quite low and for this reason it is seldom used on shoes involved in arduous wear conditions. The majority of leather soles are to be found on high-class fashion shoes. Another potential problem with leather soles is the low slip resistance which tends to manifest itself in both wet and dry conditions.

### *9.3.2 Resin Rubber*

This is by far the most common sheet soling material, and was developed 30–40 years ago primarily as a replacement for leather. Today more than half the resin rubber is used on ladies' shoes and in all applications it performs with very little trouble. It is based on SBR with high styrene resin as a reinforcement and usually has a hardness above 90 IRHD.

As with all vulcanised rubber compounds, the formulations are complex and compounding considerations can of course be related to either performance or economy. Selection of a suitable grade of resin rubber is certainly possible to fulfil all the service requirements of a soling. Even adequate durability at the normal thicknesses of 2·7 mm for ladies' and 4·3 mm for men's is easily attainable. In common with all vulcanised rubbers, it is possible for the compounding ingredients or their reaction products to migrate to the bonding surface. If this occurs, then a weak boundary layer may be formed between the applied adhesive and the actual surface of the rubber and in such cases poor adhesion would result. To obviate this situation, chemical surface preparations have been developed[11] to remove surface contaminates prior to bonding.

TABLE 3

SHEET SOLINGS: SUMMARY OF PROPERTIES

| *Property* | *Soling* | | | | | | |
|---|---|---|---|---|---|---|---|
| | *Leather* | *Crepe rubber* | *Resin rubber* | *Gristle rubber* | *Cellular gristle rubber* | *Micro-cellular rubber* | *Cross-linked EVA* |
| Durability | × | √ | √ | √ | √ | √? | √ |
| Flex crack resistance | √ | √ | √ | √ | √ | √ | √ |
| Slip (dry) | × | √ | √ | √ | √ | √ | √ |
| Slip (wet) | × | × | √ | √ | √? | √ | √ |
| Adhesion | √ | √? | √ | √ | √ | √ | √? |

√ Satisfactory performance.
? May be a problem but dependent on compound.
× Could be a problem under certain circumstances.

*9.3.3 Crepe Rubber*

Crepe rubber is of course a natural product and its use as a soling is generally dictated by fashion demands. Both plantation and home mill grades of crepe are used as solings. The durability is good, although this property can be badly affected by contamination with oil. Soles may also tend to balloon somewhat as the compression set can sometimes be high, and again this problem is accentuated by oil contamination and the effect of heat. Flex cracking is certainly not a problem, in fact soft compounds may be self-sealing. However, slip in wet conditions may present a hazard, especially with worn soles.

Another area where occasional problems occur is in adhesion. Several adhesion systems are available for crepe and it is important to select the correct one to suit the grade of rubber as all may not be equally effective.

*9.3.4 Gristle Rubber*

This is a premium soling material and is very durable. In fact it is ideal for hard-wearing applications such as sports shoes and casual walking shoes. It is based on a high-quality SBR copolymer system which is reinforced with silica filler and apart from the sheet form, this type of compound is sometimes produced as shaped units by compression moulding.

The performance as a soling is outstanding except in slip requirements on some wet surfaces, especially when the soling has worn smooth. Again the adhesion may be affected by migration of compound ingredients to the surface but, as explained previously, this effect can easily be overcome.

During recent years cellular gristle rubber soling compounds have been developed with densities between 0·4 and 0·8. Such compounds have found application as the wearing surface in lightweight casual sports shoes such as trainers. They have a good performance but, as expected, the durability is dependent on the density and the compound quality.

*9.3.5 Microcellular Rubber and EVA*

Microcellular rubbers are quite popular on casual footwear because of their lightness. However, although they have adequate performance, the durability is related to the density and a density below $0{\cdot}5\,g/cm^3$ is unadvisable.

A more recent addition in this type of market is microcellular crosslinked ethylene vinyl acetate. The relative proportions of polyethylene and vinyl acetate in the base copolymer can be varied and it can be blended with polyethylene, SBR or polyisoprene and thus a wide range of properties can be obtained. This soling material appears to have a significant growth potential as it can be produced in a wider colour range and lighter colours than the normal microcellular rubber.

Another very significant advantage over microcellular rubber is that at equivalent densities the specific durability of the crosslinked EVA can be twice that of microcellular rubber. Because of the wide range of compounding, potential adhesion may be a problem in that a single bonding system is unlikely to be satisfactory for all compounds. This therefore necessitates careful pre-testing to establish a suitable adhesive system prior to footwear production.

In view of the stability in the market and in fact the possible increase in the use of sheet solings, other compounds are being evaluated. Sheet reconstituted polyurethane is currently being used as a soling for fashion slippers and ladies' footwear as a replacement for resin rubber. Also in this same area crosslinked EVA sheet with a density of about 0·8 g/cm$^3$ is being considered, and trials have been made with thermoplastic rubber in sheet form.

## 10 MOULDED SOLINGS

### 10.1 Thermoplastic Rubber

Thermoplastic rubber compounds based on SBS (styrene–butadiene–styrene) block copolymers were introduced into the footwear industry during the late 1960s. In Europe and North America these compounds now occupy a large part of the solings market. However, in the UK, because of the high usage of vulcanised rubber and PVC in solings, the footwear manufacturers have been less inclined to use thermoplastic rubber. Currently it is used to make about 9 million pairs of soles of which about 90% are produced in the form of moulded units.

The appearance of a thermoplastic rubber soling is generally considered to be superior to that of a PVC soling. In fact it more closely resembles a vulcanised rubber soling and as a result of this has replaced some of the rubber unit market as well as the PVC market.

The flex crack performance of this type of soling is normally excellent at low temperatures. However, with some compounds an unusual temperature effect occurs in that the flex crack performance can become worse as the temperature is increased. If therefore the footwear is to be exported to countries with hot climates, it is appropriate to ensure that the compound is satisfactory in this respect. Again, with some compounds fatigue flexing may cause a reduction in hardness. This is not normally a problem in service except with some of the harder compounds used on specialised sports footwear.

The durability of thermoplastic rubber compounds is adequate for most purposes, although it was originally assumed that the durability increased with hardness. In fact with thermoplastic rubbers a whole range of compounds is available covering the hardnesses 45–90 IRHD. During the past few years significant advances in compounding technology have occurred and today durable soft compounds are certainly available.

There is, however, one area where the use of thermoplastic rubber soles is not advocated. With fast indoor sports such as squash, thermoplastic rubber soles will certainly allow one to stop and turn very quickly because of the excellent traction characteristics. Unfortunately this results in friction melting and very rapid wear. To date this problem has not been solved and vulcanised rubber solings have a distinct advantage over thermoplastic rubber under such service conditions.

Originally adhesion to thermoplastic rubber compounds presented very significant problems. A large amount of work has been done at SATRA[12] and elsewhere on this subject to overcome the difficulty. One of the most successful methods used today involves a simple surface halogenation to increase the polarity of the rubber surface prior to adhesive application. It is also possible to modify the urethane adhesive[13] to improve the specific adhesion to a compound based on the SBS copolymer.

### 10.2 PVC

Polyvinylchloride is by far the most common thermoplastic soling used in the UK. It occupies about 28% of the soling market and is split approximately equally between units and direct injection moulding. In this respect the UK industry is unique and the high usage of PVC is governed by both economics and by the fact that this soling performs adequately in our particular climatic conditions.

The compound used is plasticised to give the required flexibility and the hardness is normally in the range 58–74 IRHD. Standard compounds have adequate durability and flex crack resistance. If worn in cold climates flex cracking can become a problem, although compound modifications such as a change in the plasticiser type can improve this performance characteristic. If the compound is too hard, this soling may be slippery and it does tend to have limited slip resistance on some surfaces when wet, although a suitable wearing pattern design can improve the situation.

PVC as a soling is so well established[14] that attempts have been made in recent years to upgrade its performance with a view to extending its use as a soling still further. This can be done by compounding with nitrile rubber or thermoplastic polyurethane to improve the wear performance and particularly the oil resistance. Such concepts are still being pursued, and both Elvaloy and Hytrel have recently been used in this context. Elvaloy in particular has been widely used in PVC soles used in the USA for sports shoes. The claims made for such a blend include better aesthetics (a more rubber-like appearance) and improved traction.

**10.3 Reaction-Moulded Polyurethane**

This type of compound was introduced as a soling at about the same time as thermoplastic rubber and has shown parallel growth with current usage at about 7·5 million pairs. It is formed by reacting together either hydroxyl-terminated polyesters or polyethers with a polyfunctional isocyanate. In the UK footwear industry polyester polyurethanes are preferred as generally they have superior physical properties, although in other countries polyethers are widely used and they perform perfectly satisfactorily.

This is a very light cellular soling with a density ranging upwards from 0·5 g/cm$^3$. The surface skin is usually thin and the durability tends to increase with increase in density. The durability of some grades is excellent and this soling is used for work boots and training shoes as well as a fashion soling.

Under normal circumstances, a polyurethane soling has adequate flex crack resistance. However, because it is light it can be produced in thick sections when fashion demands this type of soling. In such cases careful control of the compound and the design of the sole are required to avoid high stress concentrations on the wearing surface which may result in premature cracking.

**10.4 Hard Sports Solings**

These are usually high-modulus compounds, and typical solings used in this potential growth area are shown in Table 4. Durability is not usually a problem area with these solings as they are used mainly on soft playing

TABLE 4
SOLINGS FOR SPORTS FOOTWEAR

| |
|---|
| Vulcanised rubber |
| Thermoplastic polyurethane |
| Reaction-moulded polyurethane |
| Nylon |
| Thermoplastic rubber |
| Surlyn |
| Arnitel |
| Hytrel |
| PVC blends |

areas such as in soccer or rugby. Alternatively, they may not be in contact with the ground at all as in ski boots. However, the type of wear may be important because if these hard solings wear to produce a sharp edge, it can be dangerous. Nylon is subject to this type of wear and because of this there is some movement away from nylon as a stud material.

With track shoes and football boots, flex cracking is not a serious problem as these sports are not allowed in icy conditions. This, however, is not so with ski boots where very cold conditions are imperative. For cross-country ski boots the temperature/flex crack requirements are such that only the Hytrel or Arnitel, vulcanised rubber and reaction-moulded polyurethane and the TR solings would be suitable.

The hard solings can be slippery on hard surfaces, although this is not a problem on playing surfaces. Sportsmen are aware of the potential for slip on hard surfaces and act accordingly.

## 11 TWO-COMPOUND MOULDING

The basic idea of using two different materials in a moulded shoe sole is not new. In the 1930s vulcanised rubber units were made in North America with an inserted tread area of higher-grade rubber which was claimed to improve the wear performance. Of course the sponge-moulded slipper soles also have an inserted wearing surface.

However, during the last five years there has been a considerable growth in this type of dual-compound soling. It is possible to produce solings with different-coloured wearing surfaces and bonding surfaces. This can add to the appeal of the footwear in certain markets and thus increase sales. Apart from the possibility of improving the aesthetics of the sole and hence the shoe, it is also possible to improve the performance by using different materials on the wearing and bonding surface, as shown in Fig. 7. Alternatively the same type of compound may be used but of differing qualities, as shown in Fig. 8.

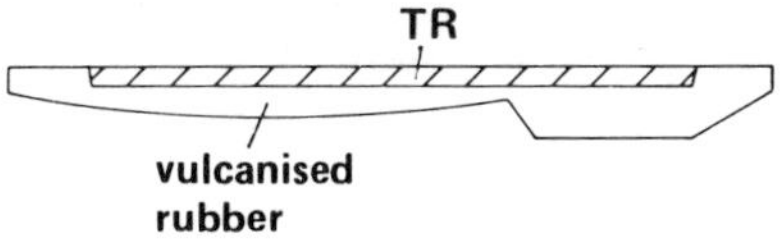


FIG. 7. Vulcanised rubber/thermoplastic rubber dual-moulded sole.

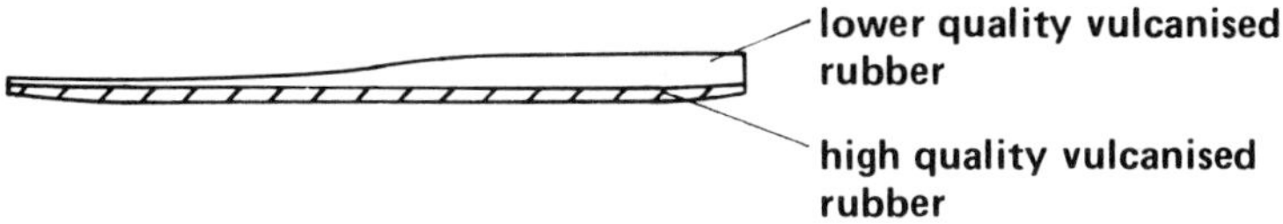


FIG. 8. Two qualities of vulcanised rubber in dual-moulded sole.

These types of solings are now being produced in thermoplastics, reaction-moulded polyurethane and vulcanised rubbers. One method of production is to use simple insert moulding as in the case of the thermoplastic rubber/vulcanised rubber combination shown in Fig. 7. In this example advantage is being taken of the high wear performance of the vulcanised rubber and at the same time the opportunity to produce a two-colour effect is available. The unit is made by moulding the wearing surface in vulcanised rubber, or cutting it from rubber sheet. The top surface is then halogenated and coated with a modified urethane adhesive before inserting in a mould and injecting on the thermoplastic rubber.

In the case of the unit shown in Fig. 8, both parts are moulded in the same mould. A basic requirement in both methods of manufacture is a high degree of compatibility between the two layers of the soling. This is not always easy, especially when using dissimilar materials, and at SATRA[15] some in-depth studies of compatibility under moulding conditions have been made.

As previously mentioned, the techniques for producing the two-compound solings are being used for various soling types. However, rather than discuss all the moulding techniques, it is intended to illustrate the complexity of the moulding equipment required by referring to some recent machines developed for moulding dual-compound solings in vulcanised rubber.

## 12 MACHINERY FOR MOULDING DUAL-COMPOUND SOLINGS IN VULCANISED RUBBER

### 12.1 Unit Machines

Both Desma and MAS produce similar machines capable of moulding two-compound vulcanised rubber units. The Desma machine is fed with an unvulcanised strip rubber which is injected using two separate injection units.

The actual mould design illustrated in Fig. 9 is quite complex. It

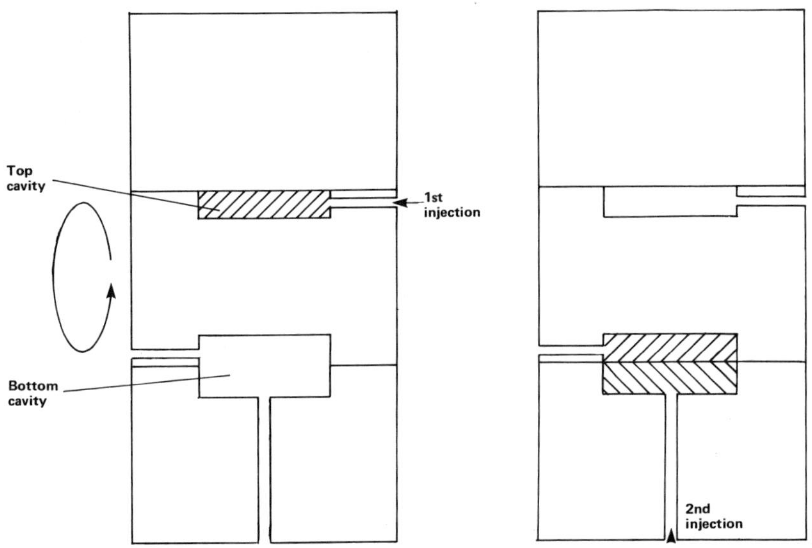


FIG. 9. Mould assembly on Desma 962 machine.

consists of three separate preheated sections with the double-headed centre section having identical cavities. The wearing surface of the unit is first injected into the top cavity and partially cured. At this stage the top plate lifts up, the bottom plate is lowered and the centre mid-plate rotates through 180° taking the partially cured wearing layer with it. The whole assembly is then reclamped and the second injection is made through the base plate to form the top surface of the sole unit. The whole unit is then fully cured and simultaneously a further wearing surface is injected into the top cavity.

## 12.2 Direct-Moulding Machines

Two machines developed during the past few years by Lucas and Nova are capable of moulding two-colour vulcanised rubber soles directly on to the shoe using compression moulding techniques.

The Lucas VO5 Bicolor machine mould assembly is shown in Fig. 10. As with normal direct moulding, the upper is positioned on a metal moulding foot and the bonding area is prepared and coated with adhesive. A rubber blank is cut to shape and placed on a false heated base plate which is automatically positioned under the moulding foot. The foot and the blank are brought into contact and the rubber is partially cured under heat and pressure and the bond to the upper is made. The false bottom plate is then lowered and the true base plate with the second colour blank is automatically slid into position. Pressure is again applied to cure fully the outsole and the already partially cured throughsole.

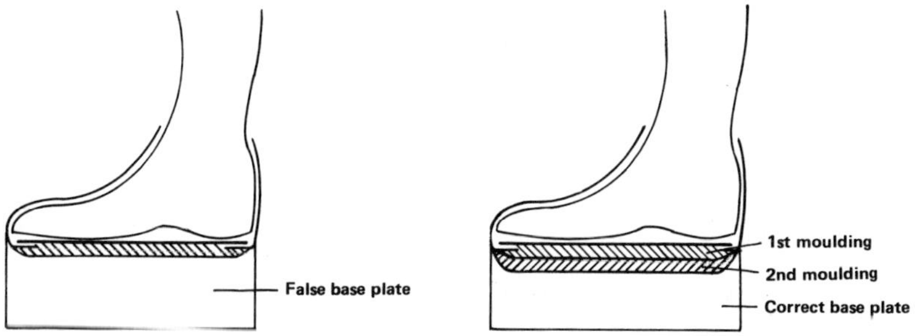


FIG. 10. Mould assembly on Lucas VO5 machine.

The Nova NPV 751B machine uses a different principle in that the wearing surface is initially precured. The mould assembly used with the Nova NPV 751B machine is shown in Fig. 11: a double-headed base plate is used initially to mould a blank of the wearing surface and partially cure it in the top cavity. The second colour blank is placed in contact with the upper mounted on a metal moulding foot in the inverted position. This foot is moved along a track to a position under the base plate which is then rotated through 180°. The partially cured wearing surface is brought into contact with the uncured blank on the moulding foot and the whole assembly is fully cured under pressure. At the same time a second wearing surface is being partially cured and another moulding foot is being made ready to accept it. The cure systems

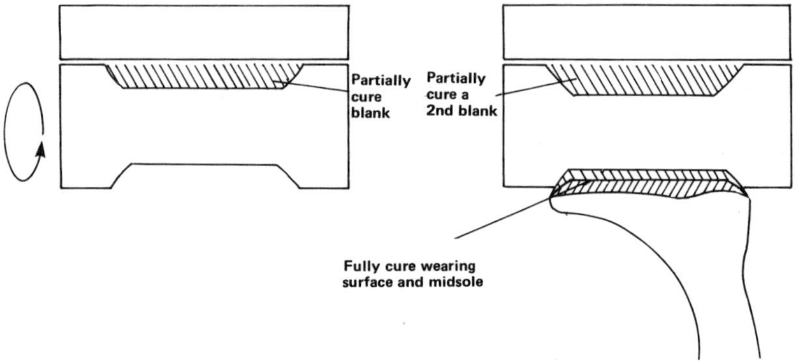


FIG. 11. Mould assembly on Nova NPV 751B machine.

used in the compounds have to be selected to enable a different cure rate in each section to be obtained.

The machinery described to attain dual-compound moulded solings is relatively complex and the processes are more expensive than in the production of single soles. Such solings, however, because of their improved aesthetics and performance are able to attract a premium price.

## 13 REQUIREMENTS OF A SHOE SOLING

The footwear manufacturer will choose a soling that will suit his particular range of styles. He will therefore be very concerned that the soling is compatible with his upper design and that the combination is acceptable to the customer. Furthermore, he will attempt to ensure that the soling compound will meet the performance requirements of the customer.

As far as the customer is concerned, there are several fundamental performance requirements. There is normally a need for the soling to flex during each step, and it is therefore necessary that the compound will undergo such a fatigue mechanism without cracking. Also during walking abrasion occurs between the soling and the ground. If the rate of abrasion is high, then the soling may wear through in a short time and require renewing. An adequate abrasion resistance is therefore necessary.

Another essential of a soling compound is a high coefficient of friction, otherwise a soling may be slippery. It is important that this property is ascertained under wet as well as dry conditions as, of course, both will prevail in practice. Slip properties are particularly important as a fall may result in personal injury. A soling should also have good dimensional stability both under the influence of load and in fact temperature. A change in dimension of the sole during service can result in unsightly footwear which may result in rejection by the purchaser.

Perhaps the most critical requirement of the shoe soling is that it should remain attached to the upper during its service life. Today most solings are attached to the upper part of the shoe using an adhesive and hence it is essential that the adhesion is adequate. It is normal and convenient for the shoe manufacturer to use a single type of adhesive for attaching the soling to the upper material. This acceptable philosophy does, however, involve some problems. In practice a urethane adhesive is used to attach the majority of the solings to the upper materials, but this adhesive is not always compatible with the soling. In view of this it may be necessary to resort to modification of the surface of the soling to ensure compatibility and hence adequate adhesion.

## 14 TEST METHODS FOR SHOE SOLING COMPOUNDS

The quality of soling compounds is established using the accepted criteria of tensile properties, hardness and density. However, SATRA has developed special additional tests for soling compounds. These tests described below have been accepted throughout the world and have been specifically designed to ascertain the practical performance of footwear solings.

### 14.1 Resistance to Flex Cracking

The flex crack performance of a soling may depend on the following factors:

(i) basic compound;
(ii) environmental temperature at which the soling is worn;
(iii) thickness of the soling;
(iv) strain exerted on the wearing surface during walking;
(v) pattern on the wearing surface.

The SATRA Ross flex test is the most common method used to assess this property and the Ross flexing machine is shown in Fig. 12. In this test a $2{\cdot}5 \times 15$ cm sample is flexed through $90^\circ$ around a mandrel at 60

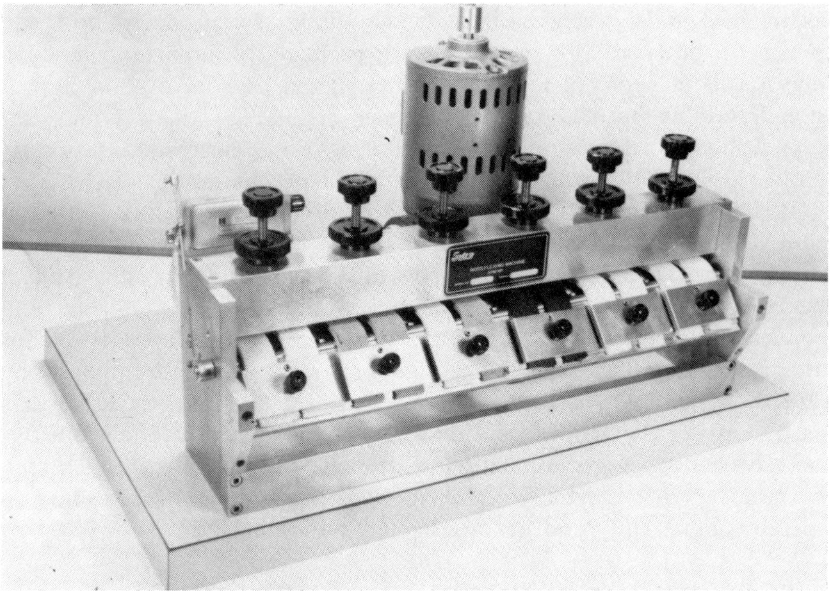

FIG. 12. SATRA Ross flex-testing machine. (Courtesy: SATRA)

flexes/min. Initially a 2 mm cut is inserted at the point of maximum flexing and the cut growth rate is measured at the required test temperature. The machine is portable and can be placed in a refrigerator for testing at low temperature.

The two major disadvantages of this test are:

1. The sample tends to heat up during flexing and may be 5–15°C above the environmental temperature.
2. The strain on the convex surface of the test piece may not necessarily be representative of the strain obtained in normal wear. However, experience with this test has enabled standards to be prepared for the common solings which correlate reasonably well with wear performance.

Provision is made to insert six samples in the machine, but all the spaces need not be filled. It is also possible to test whole shoes or parts cut from soles when the pattern on the wearing surface can be assessed.

An alternative method for testing flex crack performance is the PFI test, which was developed in Germany several years ago, and is shown in

Fig. 13. This test was initially developed specifically for testing flexible reaction-moulded polyurethane units. When sample slabs are used, the recommended thickness is 10 mm and the test piece is backed with an insole board. Flexing is again through 90° but at a rate of 150 flexes/min and, if the test piece is wide enough, three 2 mm cuts are inserted across the flexing area. The requirement is that after 30 000 cycles the cracks should not grow to greater than 6 mm. Where possible the actual unit, with the sole pattern and at full thickness, backed with an insole board, is also tested. This in fact is normal procedure with work boot units where the initial cuts are made at the base of the sole pattern.

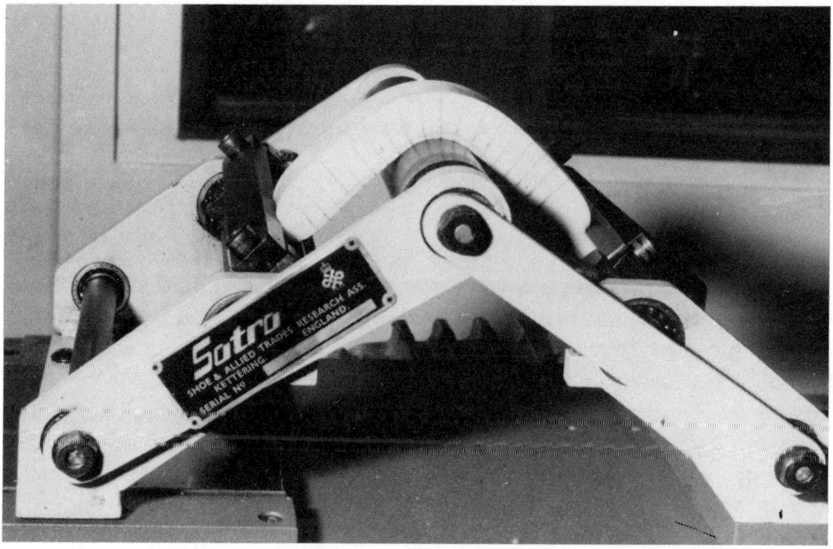


FIG. 13. PFI flex-testing machine. (Courtesy: SATRA)

### 14.2 Durability

The best method of assessing the durability of a soling is by a practical wear trial. In a full durability trial five pairs of shoes are worn, comparing an unknown top-piece or sole with a standard material in a left/right comparison. After a suitable wear period the shoes are collected and the volume loss of the unknown and standard top-pieces determined. The shoes are then reissued with new top-pieces which have been reversed with the standard to accommodate left/right wear bias. Again the volume losses are measured after the wear period and the geometric mean of the total volume losses obtained. The ratio volume loss of standard/volume loss of unknown sample is then quoted as the specific durability of the unknown soling.

A high specific durability is required for arduous wear conditions and invariably the specific durability of synthetic solings is higher than that of leather. Leather may have a specific durability of about 0·5 whereas some of the high-quality silica-filled SBR rubbers may be higher than 3·0.

Durability wear trials are time-consuming and laboratory abrasion machines are available. With non-thermoplastic materials such as vulcanised rubbers, it is sometimes possible to obtain good correlation between the abrasion test result and wear performance. However, this is not always true. Thermoplastic soling materials present problems with abrasion tests as the temperature of the test piece increases owing to friction. The properties of a thermoplastic compound change significantly with temperature and often it is not possible to obtain a reasonable correlation between rate of abrasion in an abrasion test and wear performance.

### 14.3 Slip

Although laboratory friction tests are available, it is again considered preferable to test the slip properties of a soling in a practical test. A ramp test is used in which normal walking can be done on any surface simply by changing the base material on the ramp and, of course, the angle of the ramp can be varied. The test can be done in dry or wet conditions by hosing down the walking surface. Slippery soles can, of course, cause problems as personal injury may result from a fall. Attempts are made to ensure that the solings tested in this respect will be safe for the customer. A minimum angle of 15° on the ramp is the requirement and for completeness the test is done with and without a sole patterm to simulate a new and worn sole. Of course, the sole pattern can have a great effect on the slip performance and a good anti-slip pattern should have numerous leading edges enabling traction in all directions.

### 14.4 Spread

Spreading of solings in wear, or collapse in the case of cellular soling materials, can, as mentioned previously, quickly lead to unsightly footwear. This property is assessed by the compression set tests in which a sample is compressed between two plates at a known pressure for 24 h. The pressure is then released and the change in thickness measured after 30 min for solid samples and after 1 h for cellular samples. Collapse or spread are usually not really severe problem areas as this test is so easy to perform that compounds can be readily assessed before they penetrate the footwear market.

### 14.5 Adhesion

Sole adhesion, however, is a very significant problem area and the most common cause of complaint for shoe returns from the customer. Adequate bonding systems have been developed for most soling materials.[16] When changing to a new soling material or adhesive system it must be checked by at least a laboratory test and a factory production trial, and this should preferably be coupled with a wear trial before going into full production. However, the procedures within the shoe factory can be quite critical and, assuming that the adhesive system is correct, then failures can result due to the following factors:

(i) insufficient surface preparation of sole or upper;
(ii) inadequate priming of surfaces;
(iii) too long open time on the adhesives;
(iv) insufficient heat reactivation of adhesive;
(v) insufficient pressure or uneven pressure in sole pressing.

The laboratory adhesion test involves the preparation of laminates, reproducing as closely as possible the exact conditions used in production. The laminate is then tested in the peel mode using a tensiometer with a jaw separation rate of 10 cm/min. Ideally separation would occur in a substrate, but if actual adhesion failure occurs rather than cohesion in the substrate, then guideline values are available for various assemblies to indicate whether the bond is satisfactory.


## ACKNOWLEDGEMENTS

The authors wish to thank the Shoe and Allied Trades Research Association (SATRA) for permission to publish this chapter.


## REFERENCES


1. Gillibrand, J. (1968) *J. BBSI*, **15**, 195.
2. SATRA Modern Shoemaking No 4, *Shoemaking with PUCFs.*
3. Hole, L. G., and Bingham, J. A. (1977) *SATRA Bull.*, **17**, 577.
4. Austin, P. R. (1942) US Patent 2 302 167, 17 November.
5. Payne, A. R. and Whittaker, R. E. (1970) *J. Inst. Rubb. Ind.*, **4**, 107.

6. HOLE, L. G. (1970) *Rubb. J.*, **152**, 72.
7. HOLE, L. G. and WHITTAKER, R. E. (1971) *J. Mat. Sci.*, **6**, 1.
8. WHITTAKER, R. E. (1971) *J. Appl. Polym. Sci.*, **15**, 1205.
9. WHITTAKER, R. E. (1979) Chapter 6 in *Elastomers: Criteria for Engineering Design* (ed. C. Hepburn and R. J. W. Reynolds) Applied Science Publishers, London.
10. HOLE, L. G. (1970) Chapter 6 in *Poromerics in the Shoe Industry* (ed. A. R. Payne) Applied Science Publishers, London.
11. PETTIT, D. and CARTER, A. R. (1973) *J. Adhes.*, **5**, 333.
12. WAKE, W. C. (1977) *Developments in Adhesives—1*, Applied Science Publishers, London, p. 223.
13. WEBB, C. E. (1980) *Clothing Footwear J.*, **1**(5), 354.
14. CARTER, A. R. (1979) *Plast. Rubb. Proc.*, **4**(2), 61.
15. CARTER, A. R. and DENTON, M. F. (1979) Dual compound moulding and insert moulding—compatibility studies. SATRA Report 8, December.
16. CARTER, A. R. (1978) *Eur. Rubb. J.*, **160**(1), 31.

*Chapter 3*

# THE LATEX AND ALLIED INDUSTRIES

A. Ford

*Bayer UK Limited, Altrincham, Cheshire, UK*

*SUMMARY*

*Over the last 20 years the use of synthetic latices has increased very substantially, in conjunction with NR latex, and is now superseding it in terms of volume. The most significant growth has been in the carpet backing industry. This has shown a remarkable volume increase and is by far the biggest consumer of latex. However, other interesting growths have taken place: NBR latex as a binder, CR latex as a cellulose binder, NBR and CR latices for gloves, and the latest growth area of aqueous adhesives. The foam industry is also still very important and may increase if legislation regarding flame retardancy comes about. A state-of-the-art review of polyurethane foam concludes the chapter.*

## 1 INTRODUCTION

Perhaps the most significant development since the 1950s has been the growth in the use of synthetic latices both with, and to a great extent as a replacement for, NR latex. Possibly the most significant growth area is in the carpet backing industry, particularly with the advent of the tufted carpet and the development of carpet underlay and foam backing. In the latex industry the sole use of NR latex has been updated by the introduction of synthetics and developments over recent years have completely broadened the application base:

Carpet backing:
SBR
XSBR
NR

Foam industry:
SBR–HS
NR

Paper coatings and impregnations:
XSBR for cellulose fibres
NBR for non-wovens, interlinings, as base for synthetic leather
CR for Texon board
NR for leather board
Acrylics for hygienics and non-wovens.

Dipping:
NR for gloves, balloons, condoms
NBR for gloves
CR for gloves and balloons

Binder for animal hair:
NR, CR or blends for flame resistance.

There has also been an increased interest of late in the adhesives industry because of the limitations regarding the use and extraction of solvents.

Some recent figures[1] regarding relative market sizes, in dry tonnes/year, may be of interest:

| | |
|---|---|
| SBR–HS | 28 000 |
| XSBR–HS | 5 000 |
| XSBR | 30 000 |
| NBR | 1 500 |
| CR | 2 000 |
| PV acetate | 12 000 (paint industry) |
| NR | 7 000 |
| Acrylics and others | 2 000 |
| | 85 500 |

The above figures clearly indicate the predominance of the carpet industry, which is far in advance of other sectors such as impregnations and dipping.

## 2 CARPET BACKING

The market split is as follows: pre-coat and foam backing account for 70%, secondary backing for 30%, with primary backing as a speciality. The size of the current latex market is:

| | |
|---|---|
| Pre-coat and secondary backing | 20 000 tonnes XSBR |
| Foam backing (gel/no-gel) | 18 000 tonnes SBR (HS) +some NR latex |

*Pre-coat formulation (parts by weight)*[2]

| | |
|---|---|
| Polysar Latex 793 | 100·0 |
| Calgon (25% solution) | 0·3 |
| Calcium carbonate | 300·0 |
| Water to 75% total solids | |
| Polyacrylate thickener to 10 Pa s | |

*Foam backing formulation (parts by weight)*

| | |
|---|---|
| Polysar Latex 730 | 100·0 |
| Sodium alkyl sulphosuccinamate | 5·0 |
| Calgon | 0·5 |
| Potassium hydroxide | 0·2 |
| Sulphur | 1·65 |
| Zinc diethyl dithiocarbamate (ZDC) | 1·25 |
| Zinc mercaptobenzthiazole (ZMBT) | 1·00 |
| Zinc oxide | 2·0 |
| Non-staining antioxidant | 0·75 |
| Calcium carbonate | 75, 140 or 200 |
| Water to 75% solids | |
| Pigment | 0·5 |
| Polyacrylate thickener to 5 Pa s | |

For non-carboxylated latices, crosslinking is achieved by conventional accelerators such as ZDC, MBT and ZMBT.

Antioxidants are used in all systems and are normally of the non-staining type. There has been a substantial interest of late in antioxidants of the self-emulsifying type, for ease of processing.

## 3 FOAM INDUSTRY

The foam industry is also a large consumer of latex but has been affected by the development and growth of PU foams. The latter are now used extensively in the furniture and automobile sectors. Latex foam is still used for mattress and pillow requirements and is also used as a binder of animal hair for certain qualities of car seats.

There are two main methods of producing latex foam. In one, the curable latex foam is whipped to 7–14 times the original volume to form a stable foam, then the foam is gelled by the addition of sodium silicofluoride (Dunlop process) and it remains capable of being cast for relatively long periods before rapidly gelling and coagulating.

In the Talalay process, hydrogen peroxide is added to the latex compound in which it is decomposed by adding a catalyst, e.g. yeast. The oxygen produced in this reaction expands the latex to 8–14 times its original volume. The foam is solidified by being frozen to −10 to −15 °C, after which $CO_2$ is passed through. By the time thawing takes place, the foam has already coagulated and been cured, washed and dried in the normal manner. The Talalay process gives a soft foam rubber with a particularly fine and uniform pore structure.

## 4 IMPREGNATIONS

This has been a big growth area in recent years, in particular in the use of synthetic latex as a binder for cellulose and synthetic fibres. One large area is the use of NBR latex as a binder in the impregnation of base felts for the manufacture of synthetic leather. The formulation given below is chiefly used as a binder for thick non-wovens that are split at the end of the finishing procedure.

*Non-woven reinforcement by dipping*

| | 25% | 30% | 35% | (solids content of compound) |
|---|---|---|---|---|
| (a) | 100·0 | 100·0 | 100·0 | pbw Perbunan N solids (=210 pbw Perbunan N latex 3415 M 47·5%) |
| (b) | 8·0 | 8·0 | 10·0 | pbw 25% aqueous solution of Emulvin W |
| | 100·0 | 60·0 | 15·0 | pbw water |

(c) 2·5 pbw ZnO active
1·5 pbw colloidal sulphur 95
0·8 pbw Vulkacit LDA
2·0 pbw Bayertitan R-FK-D
0·2 pbw Vulkanosol dyestuff
} dispersed in 11·0 pbw of a 5% aqueous solution of Vultamol

| | | | | |
|---|---|---|---|---|
| (d) | 8·0 | 8·0 | 10·0 | pbw 25% aqueous solution of Emulvin W |
| | 2·0 | 1·0 | 1·0 | pbw coagulant WS |
| | 106·0 | 68·0 | 64·0 | pbw water |
| | 38 °C | 38 °C | 40 °C | (coagulation point after 1 h) |

Drying is followed by vulcanisation in hot air at 110–130 °C for 20 min. A typical installation is shown in Fig. 1.

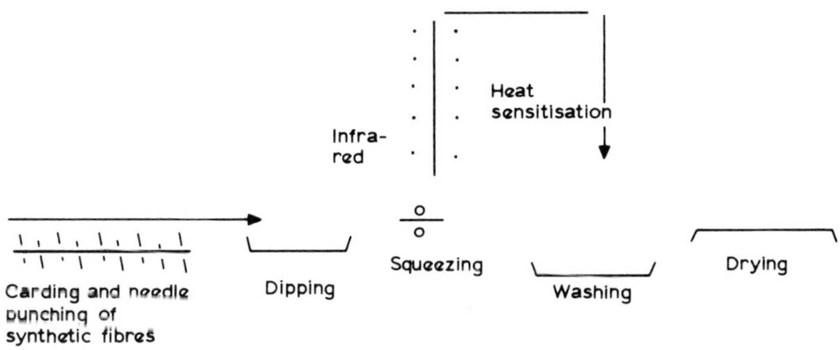


FIG. 1

The shoe and associated industries also call for interlining materials and counters, where cellulose fibres are bound with CR and NBR latices. Modern trends and technical developments mean that cost factors have necessitated blends with SBR, without too much deterioration of properties. The use of PVA is also of interest. Some grades of interlinings can be produced entirely with SBR, provided that dry cleaning is not foreseen.

*Formulation for toe caps and heel stiffeners*

(a) 100·0 pbw Baypren solid (182 pbw latex SK 55%).
(b) 1·0–2·0 pbw 20% aqueous solution of Emulvin W.
(c) 6·0 pbw Vulkanox KSM–EM 33%.
(d) 7·5 pbw ZnO active, dispersed in 15·0 pbw of a 5% aqueous solution of Vultamol.

## 5 DIPPING

Dipping is used extensively in the production of gloves and involves two major processes: (1) the heat-sensitised process and (2) the coagulant dipping process.

The production of NBR latex gloves by the heat-sensitising process may be carried out using the following guide formulation:

(a) 100·0 pbw Perbunan N solid (200 pbw Perbunan N latex T 50%)
(b) 0·5 pbw coagulant WS
 10·0 pbw 20% aqueous solution of Emulvin W
 0·25 pbw Baysilone Defoamer E
(c) 2·5 pbw ZnO active
 1·5 pbw colloidal sulphur 95
 0·8 pbw Vulkacit LDA
 2·0 pbw Bayertitan R-FK-D
 0·2 pbw Vulkanosol dyestuff
 } dispersed in 12·2 pbw of a 5% aqueous solution of Vultamol
 Coagulation point of compound: 38 °C

If a metal former is heated to 80 °C, then a dipping time of 5 s produces a film thickness of about 0·5 mm. The film produced is dried for 45 min at 70 °C and vulcanised in hot air at 90–100 °C for 30 min.

Rubber gloves are often made by superimposing an NR film on a film of Baypren, both films being produced by dipping. As well as improving the tear resistance of the gloves, this facilitates flocking. The outside of the finished glove consists of Baypren and the inside of NR, so NR latex is used for the first dipping and Baypren T for the second.

*Coagulant dipping solution*
50 pbw methyl alcohol
20 pbw acetone
10 pbw calcium chloride
20 pbw calcium nitrate tetrahydrate
0·5 pbw antistatic plasticiser KA

The following guide formulations are recommended:

(a) 100 pbw Baypren solid (=173 pbw Baypren Latex T)
(b) 2·0 pbw 20% aqueous solution of Emulvin W
 0·5 pbw Helio fast paste
 } mixed together

| | | |
|---|---|---|
| (c) | 7·5 pbw ZnO active<br>2·0 pbw Bayertitan R-FK-D<br>2·0 pbw Rhenocure CA<br>1·0 pbw Vulkacit D<br>1·0 pbw colloidal sulphur 95<br>2·0 pbw Vulkanox NKF | dispersed in 24·0 pbw of a 5% aqueous solution of Vultamol |

*Composition of NR latex compound*

| | | |
|---|---|---|
| (a) | 100 pbw natural rubber solids (=167 pbw natural latex 60%)<br>2·0 pbw 20% aqueous solution of Emulvin W | |
| (b) | 2·5 pbw ZnO active<br>1·0 pbw colloidal sulphur 95<br>0·5 pbw Vulkacit LDA<br>1·0 pbw Vulkanox BKF<br>2·0 pbw Bayertitan R-FK-D | dispersed in 11·0 pbw of a 5% aqueous solution of Vultamol |

The compounds are prepared by mixing the ingredients in the order shown. Roughened porcelain formers are recommended.

The formers are dipped first in the coagulant solution and then in the Baypren Latex T compound in which it should remain for about 1 min. As soon as the film has coagulated thoroughly, the former is dipped in natural latex compound in which it should remain for 45 s. The above dipping time gives a total film thickness of 0·5–0·7 mm. The bead can be rolled 2–3 min after the last dipping. When roughened porcelain formers are used, the gloves can be cured immediately (in hot air at 130°C for 45 min), preliminary drying being unnecessary.

Meteorological balloons were for many years produced from NR. Today, owing to the high performance requirement, they are also produced from Baypren Latex T. A heat-sensitised compound is used and a dipping process or a rotational moulding process can be used.

In the dipping process, long fluted formers are heated, and in the wet gel state the balloon is blown up to the required diameter. The neck has to be added as a separate step. For the moulding process, a suitable mould is placed in a hot-water bath and, using excess compound with a combination of time and temperature, a wet gel is built up of the required thickness. After the excess compound has been removed, the wet gel is expanded to the required diameter.

The formulations used are typical for a heat-sensitisation process, with special attention being paid to plasticisation and antiozonants which are vital for good flight performance, night and day, at the very low temperatures encountered. When designing the compounds, accelerator amounts are chosen to ensure a low modulus, which is desirable in meteorological balloons. For the same reason the cure is discontinued after 20 min, i.e. before it reaches completion.

Very recently there has been a renewed interest in the production of rubberised hair for car seats, and a US standard MVSS 302 is under consideration. As greater flame retardancy is required, the following formulation may be of interest:

| | | *Wet* | *Dry* |
|---|---|---|---|
| Baypren Latex MKB 58% | | 78·0 | 45·0 |
| NR latex 60% | | 92·0 | 55·0 |
| Emulvin W 20% | | 5·0 | 1·0 |
| ZnO active | 5·5 | | |
| Vulkacit NPV/C | 0·7 | | |
| Vulkacit LDA | 0·6 | | |
| Colloidal sulphur | 1·8 | | |
| Kaoline P | 15·0 | 120·0 | 56·0 |
| Silitin Z | 10·0 | | |
| Antimony trioxide | 18·0 | | |
| Vulkanox HS powder | 1·0 | | |
| Vultamol 100% | 3·4 | | |
| Water | 64·0 | | |
| Arubren CP EM 55% | | 30·0 | 16·5 |
| Mobilcer C 50% | | 8·0 | 4·0 |
| | | 333·0 | 177·5 |

## 6 ADHESIVES

Because of increasing restrictions in the field of solvents, aqueous-based adhesives are being increasingly investigated. Owing to the very strong crystallisation of the polymer, films produced with Baypren Latex SK have high cohesion strength. It is advisable to improve the adhesion strength of the adhesives by adding suitable resins in emulsified form. Among those frequently used are terpane and terpene–phenolic resins, rosin and pentaerythrite esters. Coumarone–indene resins can also be used.

*Adhesive with moderately long open time (about 60 min):*

(a) 100 pbw Baypren solids (=182 pbw Baypren Latex SK 55%) 3·0–5·0 pbw 20% aqueous solution of Emulvin W.

(b) 6·0 pbw Vulkanox KSM–EM 33%.
(c) 7·5 pbw ZnO active, dispersed in 15·0 pbw of a 5% aqueous solution of Vultamol.
(d) 100·0 pbw Durez Resin 12603 emulsion 25%.

*Adhesive with very long open time (about 24 h):*
(a) 100·0 pbw Baypren solid (182·0 Baypren Latex SK 55%)
(b) 2·0–5·0 pbw 20% aqueous solution of Emulvin W.
(c) 6·0 pbw Vulkanox KSM–EM 33%.
(d) 7·5 pbw ZnO active, dispersed in 15·0 pbw of a 5% aqueous solution of Vultamol.
(e) 100·0 pbw 30% emulsion of Ester Gum 8L.

Experiments have been carried out lately using latex as a modifier for bitumen. This is felt to have application for sealing tunnel walls and as a roofing membrane system. Both SBR latex and CR latex can be used, depending on final property requirements.

The foregoing has indicated trends and patterns, both immediate and for the future. Meanwhile the existing and fairly recent state of the art, it is considered, will also be of interest and is therefore included in this review. Furthermore, while polyurethane is strictly speaking not a latex, its use is nevertheless firmly established within the foam sector of the industry and it is therefore considered pertinent to include this material where relevant.

## 7 LIQUID RUBBERS

In addition to the increasing interest in aqueous adhesive systems, the possible use of true liquid rubbers is also very active in adhesives;[3,4] these of course are non-solvent. Further areas of interest include the general rubber goods field, especially hoses and the proofing industry section where solventless doughs are now possible. This will ultimately lead to the future non-requirement of costly, health- and fire-hazardous solvent recovery systems, and will be major breakthroughs both in environmental conditions and processing technology.

## 8 POLYURETHANE

The use of polyurethane foam has continued to grow, and further growth in this area will continue hand in glove with fire and smoke legislation requirements as they are introduced.

Foam in polyurethane can be either flexible or rigid as required and can exist chemically as either polyether or polyester varieties. The chemical structures are of course very complex, and indeed because of the infinite permutations possible are really impossible to illustrate by formulae on paper. Because of this, the examples given below are only very simple approximations.

Polymers containing ester groups, e.g. urethane elastomers and ethylene–vinyl acetate copolymers, tend to be attacked by moisture, water, acids, bases and possibly high temperatures. The damage suffered can be considerably reduced, and thus the durability of the ultimate products increased, by adding substances such as polycarbodiimides.[5]

**8.1 Flexible Foam**

There are two main types of PU foams, those based on polyesters and those on polyethers. All polyesters of any commercial importance have their origins with diethylene glycol and adipic acid and are partially branched by the incorporation of small quantities of triol.

*Polyester reaction*

$$\underset{\text{Diethylene glycol}}{HO{-}CH_2{-}CH_2{-}O{-}CH_2{-}CH_2{-}OH} + \underset{\text{Adipic acid}}{n HOOC(CH_2)_4COOH}$$

$$\downarrow$$

$$HO{-}CH_2{-}CH_2{-}O{-}CH_2{-}CH[{-}O{-}\overset{\overset{\displaystyle O}{\|}}{C}(CH_2)_4{-}\overset{\overset{\displaystyle O}{\|}}{C}{-}O{-}CH_2{-}$$

$${-}CH_2{-}O{-}CH_2{-}CH_2]_n{-}OH \quad + 2nH_2O$$

Polyester

The reaction is carried out in stainless steel plant in an inert atmosphere, usually at a temperature of approximately 225 °C. The water liberated during the reaction is removed via a fractionating column and, upon completion of the reaction, the polyester is cooled and filtered. The polymerisation rate is checked and controlled by balancing the viscosity increase rate against the reduction in the free acid, this latter of necessity being of a high state of purity, especially from the presence of iron and other heavy metals, which must be absent to ensure batch-to-batch uniformity of the foam reaction.

Tests are then carried out to ensure the necessary control over the molecular weight, and these include colour, density, acid value, viscosity and saponification values.

The heat liberated due to the reaction of the polyester with isocyanate, and thus the temperature rise and actual rate of reaction, increases with increasing hydroxyl value, i.e. decreasing molecular weight. On the other hand, the acid end-groups retard the reaction in the lower acid value range which is usually encountered.

As mentioned above, the basic chemistry of isocyanates and its reactions, upon which the PU industry depends, is very complex, and this is outlined and excellently described by Johnson.[6] Tolylene diisocyanate (TDI) is widely used in this context.

Polyurethanes are mainly based on the interaction of isocyanates with hydroxyl-ended polyesters and polyethers and the control of these reactions is of paramount importance. Control can be achieved in three ways: (1) the choice of the appropriate temperature band; (2) the use of either a catalyst or a retarder as appropriate; (3) the use of a masked or hindered isocyanate group rather than free isocyanate groupings. Once again, reference to Johnson[6] is advised.

*Polyether reaction*

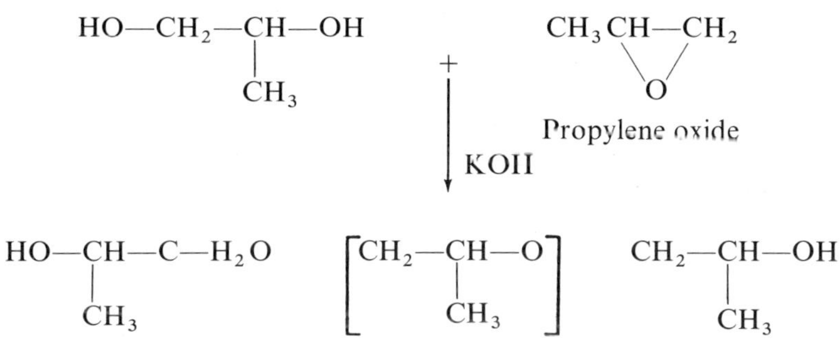


*Linear polyether*

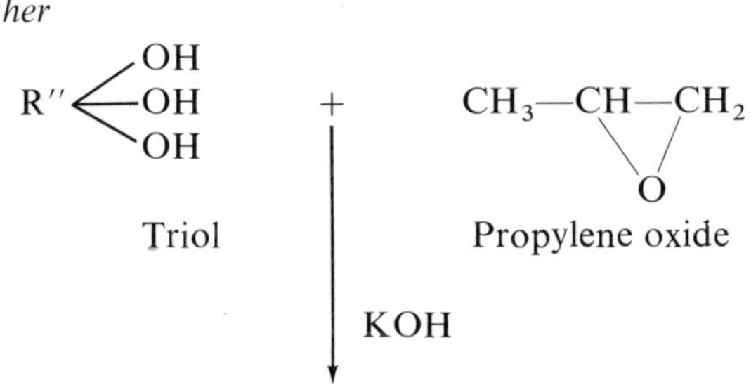

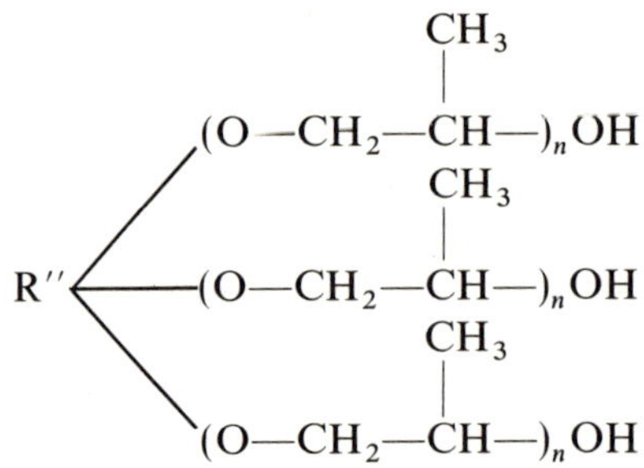

### 8.2 Branched Polyether

As can be seen, if a glycol-based initiator is used, a linear polyether is produced, whereas with a polyol such as glycerol a branched polyether is produced.

The old so-called prepolymer, which is a two-stage technique used for the manufacture of flexible foams, has long been replaced by the one-shot process currently used commercially. However, examination of the chemistry of the two-stage technique will simplify the one-shot system:

Polyether (hydroxyl ended) + Diisocyanate

↓

Polyether urethane (isocyanate ended) + Excess diisocyanate

↓ $H_2O$

Polyether–urethane urea + $CO_2$ (gas)

↓ Exotherm heat + catalysts

Additional crosslinks (Biuret formation, allophanate formation, isocyanate formation).

↓ Excess heat

Degradation + Carbodiimide formation + $CO_2$

All the above chemical reactions must be under strict control in order to produce a uniform foam cell structure.

The one-shot system consists of mixing together all the components and thus it is essential that a balance between the polymer-forming and gas-forming reactions is achieved so as to produce stable and strong cell walls capable of holding the $CO_2$ during the early stages of the reaction.

The obvious and fundamental difference between polyether- and polyester-based foams is the lower resilience of the polyester materials. This has led to the almost complete replacement of polyester in general upholstery application by polyether foams, but the polyester, which is a little more expensive, continues to be used where its unique properties are required over those of polyether, e.g. where stiffness and high tensile properties are required for such applications as luggage and handbag linings in conjunction with PVC sheet, shoulder pads for jackets and coats, and also in the footwear industry for slipper linings. Polyester foams are also ideal for insulating hot-water pipes, etc.

It is also vitally important to consider and design carefully the processing plant and equipment which is used in the PU industry. This is of equal importance to the choice of materials used in foaming and also the recipes and formulations used. Each is equally interdependent and failure to realise this will produce serious difficulties.

## 8.3 Rigid Foams

As the name implies, the essential requirement is that of *stiffness* of the polymer network so that overall rigidity is imparted to the foam. This is primarily achieved by a high degree of crosslinking, plus the choice of molecular weight, together with the functionability of either the polyester or polyether resin used during the isocyanate coupling. In the case of the polyether resin the use of a polyfunctional initiator achieves this as a basic property of these materials. Once again the stiffness can be influenced by the choice of an aromatic rather than an aliphatic nucleus on which to initiate the polyether, e.g. tolylene diamine can be used as a nucleus for the propylene oxide addition:

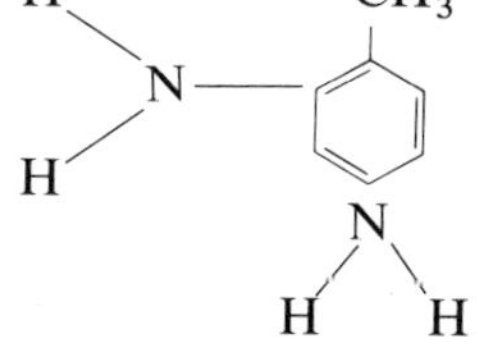


2, 4-Tolylene diamine

Additionally, it is also possible to impart flame retardance or char formation properties by selection of the necessary initiator. By the use of phosphoric acid as a hydroxyphosphorus initiator, a polyether may be produced by reacting with propylene oxide having the following composition:

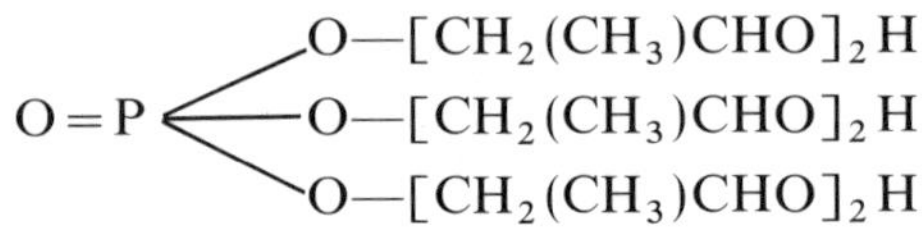

Another method of imparting further flame resistance is by the use of low-boiling halogenated alkanes as blowing agents. One such widely used material is $CCl_3F$; this is chemically inert and so there is no isocyanate consumption during the gas generation as there is with $CO_2$ blowing. These chlorofluorocarbon gas bubbles after the volatilisation and expansion during the —OH/—NCO urethane-forming reaction finally become trapped in the closed cells of the final rigid foam. $CCl_3F$ has a further extremely useful property and that is its capability of considerably reducing the viscosity of very high viscosity resins, thus making these usable for foam making.

Rigid PU foams find application and uses mainly as insulating materials in many industries including agriculture, aircraft, automobiles, ships, caravans, food containers and insulated transport, packing, refrigeration, footwear and thermal insulation. Rigid foam was initially rather slow to develop, primarily because of over-optimism at the laboratory stage, as factory processing was much more difficult to achieve than early laboratory work suggested, but its use is very widespread and still growing.

Flammability risks have to be considered for foam applications, especially where the foam is to be used in furniture and upholstery. Improvements have included the use of a cellular elastomer possessing flexibility and flame retardation properties for use as an interliner in furniture upholstery applications.[7] Indeed, in several countries legislation is already being introduced to make such a system and similar ones mandatory for safety reasons.

**8.4 Liquid Castable Polyurethanes**

In some applications liquid castable urethane elastomers based on 4,4′-diphenyl methane diisocyanate (MDI) with the diol hydroquinone di($\beta$-hydroxyethyl)ether (HQEE) have been found to give improved polymer behaviour in respect of cure rates and product performance.[8]

## REFERENCES

1. Internal Bayer UK study.
2. POLYSAR LTD (1978) International Polymer Latex Conference.
3. EVANS, C. W. (1981) International Rubber Conference, Harrogate.
4. Liquid NBR, Polysar Ltd.
5. Stabaxol PCD, Bayer AG.
6. JOHNSON, P. C. (1968) Isocyanate reactions. In *Advances in Polyurethane Technology* (ed. J. M. Buist and H. Gudgeon). Applied Science Publishers, London, Chap. 1.
7. DU PONT DE NEMOURS INC. (1977) *Du Pont Mag.*, **71**(4), 18–19.
8. HAMPSON, F. W. (1980) in *Developments in Rubber and Rubber Composites—1* (ed. C. W. Evans) Applied Science Publishers, London, p. 86.

*Chapter 4*

# HEALTH AND SAFETY CONSIDERATIONS IN RUBBER PROCESSING

G. LAWSON

*Rubber and Plastics Research Association of Great Britain, Shawbury, Shropshire, UK*

*SUMMARY*

*Industrial health and safety is of international concern but unfortunately there are few internationally agreed standards or codes of practice. In order to avoid citing information which may well be misleading or even inaccurate, depending on the nationality of the reader, the following chapter has been written specifically at the general interest level. The approach taken has been to provide a historical perspective of the many factors which can affect the safe operation of an industrial plant. Examples of the increasing legislative approach to problems concerning health and safety have been chosen from several countries to provide an international viewpoint. This chapter traces the changes in emphasis in working practices from considerations of such fundamentals as heat and light to the possible problems posed by the introduction of high technology, for example the use of lasers and radiation curing. Despite the introduction of legislation and mechanical and electrical safeguards, safe operation must ultimately rely on human vigilance to ensure that correct procedures are followed.*

## 1 INTRODUCTION

The healthy and safe operation of any industrial process requires that attention be paid to a myriad of details and regulations. Such attention is the concern, to a greater or lesser extent, of every single person on a factory site. Many of the legal requirements have been in force for years and, as such, are taken for granted. It is, however, often stated that continuing vigilance is the necessary price of safe operation and thus a certain degree of repetition has to be accepted.

In general, regulatory authorities charge management with the responsibility of providing a reasonably safe and healthy working environment. Such efforts are further spurred on by both humanitarian and economic considerations. Accidents hurt individual people and their families and may also damage the morale and efficiency of working teams. Lost-time accidents may also mean loss of money to both the employee and the employer.

Not all the onus is, however, placed on management. For example, in the UK the 1961 Factories Act requires that employees shall not wilfully misuse safety equipment and will use such equipment during operations for which it is supplied. Furthermore it is required that employees shall not wilfully do anything likely to endanger themselves or others.

This chapter has been written with the general reader in mind and does not, therefore, cover in depth all the areas relating to the safe operation of a complex industrial processing plant. Discussions are restricted to those areas which are more specific to rubber processing. Industrial safety is discussed both more generally and in greater depth in many readily available texts.[1-3]

Rubber processing is a world-wide occupation and although many health and safety considerations are internationally accepted, specific regulations and recommendations may differ from one country to another. Bearing such possibilities in mind, attempts have been made, where possible, to discuss trends rather than specific examples.

## 2 HISTORICAL PERSPECTIVE

Historically, occupational safety and health has been concerned with the promotion of optimal working conditions and hence productivity. Attaining such a goal requires the identification of those factors depriving the workforce of such conditions and the implementation of corrective measures.

The identification of the hazard is the vital stage. In many instances the hazard has been obvious—unguarded flywheels for example. Industry has perhaps reached the situation in which all the obvious hazards have been recognised and dealt with appropriately and is now having to react to situations which were previously thought to be satisfactory but which may now be open to criticism.

More advances in science and in health and safety-related legislation have been made in the past 70 years than in all prior history. As the science of medical diagnosis and the complexity of industrial society advances, it is likely that new hazards will be created, identified and regulated at an ever-increasing rate.

Information and legislation with respect to occupational safety and health should not be considered a product of this modern 'enlightened' age. Pliny the Elder (AD 1) suggested the use of dust masks when grinding ores. In 1315 King Edward III of England tried (unsuccessfully) to prohibit the burning of sea coal because of its noxious odour. During the Middle Ages problems associated with the production and use of heavy metals (lead and mercury) were defined. Ramazzinni (1700) examined the conditions of work and related diseases of many occupations of his time. Scrotal cancer among chimney sweeps, the first identified occupational cancer, was recorded by Percival Pott in 1775. These are just a few examples of early information relating to occupational safety and health.

The first major worker-health-related legislation was the UK Factory and Health Laws (*ca.* 1891). By 1909, 21 States in the USA had enacted laws attempting to regulate industrial working conditions in one way or another. Since this period similar and more legislation has been introduced in countries throughout the world. The Walsh–Healy Act (1934) in the USA was the first piece of broad legislation which included numerical occupational health standards, the Threshold Limit Values (TLV).

## 3 THE EXTENT OF CURRENT LEGISLATION

Perhaps the most significant individual items of legislation were promulgated in the early 1970s, the Occupational Safety and Health Act, 1970 (USA), the Health and Safety at Work Act, 1974 (UK), and equivalent legislation in other countries. The introduction of these Acts brought virtually every occupation and industry under the control of a regulatory authority.

The regulatory control over manufacturing industry does not end at a particular factory boundary. To an increasing extent the level of emissions from stacks is being monitored and control permits or maximum emission levels issued. In Australia for example it may be necessary, in some States, to obtain a permit before the construction of a stack is started.

Other waste materials, particularly polluted water, are controlled to various degrees depending on the country being considered. In the USA the Environmental Protection Agency has published a list of 128 priority pollutants which will be monitored in effluent and river systems. In the UK effluent emission is controlled by Water Authorities in consultation with the specific industry concerned. Thus it can be seen that health and safety considerations extend not only to the immediate workforce but also to the local neighbourhood and perhaps beyond that.

The final area of legislative control defines the nature of the materials that can be used depending on the designated use of the end-product. Such control is very important to the rubber industry, particularly when the reactive nature of many of the chemicals used in a formulation is considered. The German Federal Republic and US Food and Drug Administration issue lists of chemicals which have been considered for use in rubbers in food contact situations. Such concern has been highlighted by a recently proposed German decree specifying the maximum allowable concentration of nitrosamines in the teats for babies' bottles.

## 4 FACTORS AFFECTING INDUSTRIAL SAFETY AND HEALTH

There is a wide range of factors (Table 1) on which the safe operation of any process depends. As has been stated earlier, many of the following requirements are taken for granted. The reader is doubtless aware of the problems which can occur as a result of the absence of light during a power failure. In many of these areas the potential hazards are well recognised and adequate safeguards are and have been available for many years. In such instances only a brief discussion of the general principles involved will be given. Rather more attention will be paid to those areas which are either relatively new to the industry or where previously unsuspected hazards are now being identified.

### 4.1 Light

Lighting is taken for granted. Yet, according to Ragsdale,[4] in this supposedly technological age much of the electric lighting installed over

TABLE 1
SUMMARY OF MAJOR CATEGORIES OF FACTORS AFFECTING SAFE FACTORY OPERATION

| | |
|---|---|
| 4.1 | Light |
| 4.2 | Heat |
| 4.3 | Fire/explosion |
| 4.4 | Noise |
| 4.5 | Electrical hazards |
| 4.6 | Mechanical hazards |
| 4.7 | Radiation hazards |
| 4.8 | Dust |
| 4.9 | Chemicals: toxic<br>flammable<br>corrosive |
| 4.10 | Operator hygiene |
| 4.11 | Labelling |

the shop floor and in the offices of industry can be described as a handicap rather than an aid to safety. The hazards which can occur from poor installations are insufficient light, shadows and glare. In the UK the IES Code of Practice for Interior Lighting[5] is taken as the usual reference work, but other information is available.[6]

## 4.2 Heat

Adequate control of the temperature of the working environment can prevent many potentially hazardous situations, and may also prevent the production of poor-quality products. During cold weather the author has, on several occasions, heard the comment: 'Oh, we turn the line temperature up to warm the area.' Such action may have quite dramatic effects on the safety of such an operation. The severity of a burn may be increased and also the nature of the fumes emitted may change. It is also possible to envisage the situation in which it is too hot. Under such circumstances there may well be a reluctance to wear heavy protective clothing and thus the potential for burns again increases.

Rubber vulcanisation requires temperatures ranging typically from 150 to 250°C and thus in curing areas there is a considerable potential for burns to occur. This potential can be minimised by the provision of suitable lagging, and regular maintenance can prevent accidents caused by leakage of steam, hot gases and hot liquids.

The use of molten salt baths and fluidised beds is common in the industry and these are now generally totally enclosed in order to provide suitable safety barriers. There is also mechanical input and take-off from such equipment and thus little or no requirement for an operator to handle hot material. Such is not the case where press cures are concerned. In general an operator is required to remove the finished item from one or other of the platens. The absence of suitable protective clothing would make a severe burn almost a certainty.

Autoclave curing requires both that pressure relief valves and final venting valves are in good working order. It may also be necessary to reduce the internal pressure of the autoclave to flush out residual solvent, etc., prior to removal of the finished product.

A possible problem with air bags, in the tyre industry, can be included here. These degrade internally and it is possible for the valve to become blocked during deflation. When the press opens the bag may burst. Protective shields may be used but in general the operative's position, remote from the press, will safeguard against injuries.

### 4.3 Fire/Explosions

Fires cost the UK industry approximately £1·2 million daily. Despite the specific regulations pertaining to the prevention of fires in factories, both in the UK and elsewhere, similar financial losses occur on a world-wide basis. The lesson is obvious: fires cost money!

The rubber industry uses a considerable quantity of highly flammable liquids, e.g. petroleum-based solvents. The storage and use of such solvents is generally licensed by the local regulating authority. Use on the shop floor is not so easily controlled, however. Certain conditions must be observed: the use of spill-proof containers, the prohibition of smoking and naked lights, the use of spark-proof electrical equipment and the provision of good ventilation to prevent the build-up of flammable solvent concentrations. The danger of static electricity build-up, particularly applicable in rubber factories, must be prevented by ensuring adequate earthing for equipment and personnel. The introduction of bulk powder-handling facilities also requires that careful attention be given to the potential build-up of static electricity.

Certain materials are prone to spontaneous combustion: oily rags or oil mixed with sawdust will spontaneously ignite within hours if left in a bin. Heat is used widely in the rubber industry and finished products are often left to cool down. The close proximity of flammable waste and cooling products has been one of the chief sources of fires in the rubber industry. It is therefore essential that specific instructions be given concerning the handling of all materials and that the standards of housekeeping should be exemplary.

The possibility of static electricity leading to either a fire or an explosion in bulk dry powder-handling facilities has already been mentioned. The possibility of such occurrences is uncertain owing to a lack of sufficient information, but many installations have been designed to incorporate suitable baffles and doors. One area within the industry in which dry powder explosions are possible is bulk collection equipment used to retain the powder from rubber buffing. Such material is specified, in the UK,[7] to be of a potentially explosive nature. Unfortunately there is no internationally accepted test for explosivity; however, an excellent résumé of the current situation is given by Palmer.[8]

### 4.4 Noise Control

There are very few areas within the polymer industry in which noise control, to meet statutory/recommended levels,[9,10] may be necessary. Some of these areas can be identified as braiding shops, regrind areas in which flail mills are utilised, and pelletisation areas. Protective ear muffs or plugs are readily available, as is information on methods of reducing noise from various sources.[11-13] Some comparisons of noise levels for various industrial operations in the USA are given in Table 2.

TABLE 2

| *Operation/area* | *Noise level (dBA)* |
|---|---|
| Wood chipping room | 109 |
| Plastic pelletiser | 102 |
| Polyethylene extruder | 101 (*sic*) |
| Rubber mill lines | 91 |
| Tyre curing area | 85 |
| Private residence | 25–50 |

### 4.5 Electricity

The use of electricity is common to the home, office and factory and thus there is more likelihood of bare wires, defective parts, and possibly more amateur electricians willing to try to correct any faults. The safe use of electricity depends on, at least, the following conditions:

(a) All installations, repairs and withdrawals should be made by qualified electricians.

(b) Broken or defective equipment should be reported immediately *and* steps taken to protect other people from such equipment.

(c) All portable tools should be checked regularly and issued only by an Electrical Department.

(d) Switchboards and other similar equipment should be kept clear of obstructions at all times.
(e) Operatives should know how to switch off the power at least to the machine under their supervision.

In dealing with electricity, the simple and safe approach is to get an expert.

### 4.6 Mechanical Hazards

Certain precautions must be observed with all machinery. Adequate training must be given; stopping devices, interlocks and guards must all be readily accessible and should be tested at the start of each shift. Cleaning and repairs should not be carried out whilst a machine is in normal motion. Repairs are usually carried out only when the main power fuse has been removed in order to prevent inadvertent starting.

It would take several volumes to describe all the different guards and safety devices associated with the wide range of machines used in the rubber industry. The types of guard may be classified as follows:

(a) *Fixed guard:* has no moving parts associated with, or dependent upon, the mechanism of any machinery and which, when in position, prevents access to a danger point or area.
(b) *Adjustable guard:* a fixed guard incorporating an adjustable element which, once adjusted, remains in that position.
(c) *Distance guard:* a guard which does not completely enclose a danger point or area but places it out of normal reach.
(d) *Interlocking guard:* a guard which has a movable part so connected to the machinery controls that (1) the part of the machine causing danger cannot be set in motion until the guard is closed, and (2) access to the danger point or area is denied while the danger exists.
(e) *Automatic guard:* a guard which is associated with, and dependent upon, the mechanism of the machinery and operates so as to remove physically from the danger area any part of a person exposed to danger.

Some machines in the rubber industry have been selected for special study and therefore merit specific inclusion. Two-roll mills are equipped with a Lunn safety barrier which is activated by involuntary movement towards the nip causing the machine to stop. The barrier should be adjusted to prevent the operator being able to reach beyond the safety limit indicated by an arrow on the side of the frame. Small laboratory mills are equipped with guards physically preventing access to the nip whilst the mill is in motion. Buffing and rasping machines require the use of both a suitable guard for the wheel and, in general, eye protection. Punching machines usually have one of two types of protection: a lifting gate which prevents access to the punch, or dual hand control which keeps both hands away from the danger area. The basic requirement on all presses is that access to the platens be denied when the press is in operation. This is generally accomplished by the use of interlocking guards. There is a wealth of information available concerning the safeguarding of machinery in BS 5304.[14]

## 4.7 Radiation Hazards

Radiation hazards are conventionally subdivided into

| | |
|---|---|
| ionising | (from radioactive isotopes, accelerators, etc.) and |
| non-ionising | (from microwave and laser sources). |

This section introduces the concept of new hazards resulting from the introduction of new technology.

### *4.7.1 Ionising Radiation*

International guidelines for radiation protection are set by the ICRP.[15] Ionising radiation is used in limited areas within the polymer industry, the most frequently encountered use being the 'thickness gauge'. This usually entails a closed radioactive source and a detector. Adequate warning notices are mandatory but in general such devices, when handled correctly, pose few problems owing to the limited power of the radioactive source.

Electron beams, used for crosslinking, are a much more powerful source of ionising radiation and in fact exceed the minimum dose required for biological change by several orders of magnitude. Such equipment is designed to provide an intense beam of electrons with sufficient energy to penetrate several millimetres of rubber. As such, when the beam is operating, care must be taken to provide adequate safety controls. Unlike a radioactive source, however, the electron beam can be switched off. In general it is recommended that exposure to ionising radiation be minimised.[16]

### *4.7.2 Non-Ionising Radiation*

Similar technological advances have introduced possible hazards in the area of non-ionising radiation. In particular, both microwave and laser technology are to be considered. At present there is little information concerning the use of microwaves, which can now be found on curing lines in many areas of the industry. Exposure levels of less than 10 mW (300 MHz–300 GHz) per $cm^2$ for an 8 h working day are considered to be acceptable within the USA.[17]

Depending on the power output, lasers can range from the purely decorative to the lethal. Doubtless the demonstration in which a hole is punched through a piece of steel using a laser is well known. At present the use of lasers within the rubber industry is not widespread, but already in the UK several companies are known to be using such techniques for finishing fabricated items. A laser sufficiently powerful to cut rubber can present a hazard to the operator. Two types of protection are required in the use of such equipment:

(a) an interlock system such that the laser cannot be operated until the workpiece is enclosed;
(b) means to ensure that transmitted or reflected radiation cannot be observed above a certain level (laser beams can travel through air) in order to prevent eye damage.

Information pertinent to suitable protection which should be installed in conjunction with laser equipment can be obtained from most regulatory authority offices.[17]

## 4.8 Dust

Some of the possible hazards associated with exposure to particulate materials have been highlighted in the recent past. For many years coal miners exposed to high levels of dust ran the risk of developing 'black lung' or other respiratory problems. The catastrophic effects of exposure to high levels of asbestos fibres (dust) have been widely published more recently.

Of more direct concern to the rubber industry is the difference of opinion, reported in the literature, concerning the effects of exposure to carbon black.[18-20] Results from the USA indicate no long-term debilitating effects whereas a somewhat similar study carried out in Russia was not so reassuring. The significance of the differences between these reports cannot readily be determined because of the difficulties involved in comparing the exact nature of both studies.

Possibly as a result of such information, and in a concerted action to improve working conditions, efficiency and to cut down on wastage, the rubber industry has dramatically reduced the exposure of its workforce to dust of all forms. Not too many years ago compounding areas used to be referred to as 'black holes', and a very apt description it was too! Such is not generally the case today, and it speaks volumes for the industry that such changes have taken place so rapidly.

The improvements in dust control have been brought about by the use of one or more of the following techniques:

(a) Confined weighing-out operations, in which the materials are manually dispensed, from hoppers, into polyethylene bags and the complete formulation, often including the bag, is loaded into the mixer.
(b) The use of fully automatic weighing-out facilities, in which many of the compounding ingredients are automatically bagged or are delivered pneumatically to the mixer. Such facilities are in general appropriate only to major manufacturers.
(c) The growth in the use of additives in dust-reduced form. These may be available in pellets, oil-bound powders or predispersions in a polymer matrix. A fairly comprehensive list of the materials available in these forms is available from RAPRA.[21]
(d) The development within the industry of the large-scale use of ventilation techniques.

## 4.9 Chemical Hazards

It was stated earlier in this chapter that the necessary step to ensure safety was 'to identify the nature of the hazard'. Nowhere is this more true than when considering a possible chemical hazard.

Hazards from chemicals can be roughly categorised as flammability, explosive, corrosive and toxic. Of these categories the first three can be readily demonstrated and are well known. Provided suitable information, in terms of handling instructions, is made available, the potential hazards can be fairly readily controlled.

The toxicological properties of many chemicals is a far more emotive area and is certainly less well defined. In general an excess of most chemicals will prove to be harmful (it is possible to drown in several inches of water). Chemicals are, however, necessary to modern industry and thus the nature of the action of many chemicals needs to be understood. In general the toxicological action can be classified either as acute or chronic.

In the acute situation the effects of a single exposure lasting seconds, minutes or hours may range from undetectable to a severity sufficient to threaten life or cause permanent impairment. The one advantage, if it may be called that, of an acute effect is that it is rapidly observable and corrective action can be taken once the hazard is identified. Chronic effects, in contrast, may be apparent only after weeks, months or years of exposure to a dosage level which is too low to demonstrate acute effects. In the long term the effects of such exposure may be more severe and may lead to cumulative poisoning (lead) or to occupational cancer.

In order to determine the relative acute toxicity of a specified series of chemicals, there is currently no recourse other than to animal experiments. The chemicals are administered in increasing doses until a level is reached which is fatal to 50% of the test species; this is termed the $LD_{50}$ value. The $LD_{50}$ value, expressed as milligrams of chemical per body weight in kilograms, was chosen by the EEC in 1967 to categorise the relative toxicity of chemicals (Table 3). It is generally accepted that chemicals with an $LD_{50}$ greater than 2000 mg/kg are unlikely to present any acute toxic hazards.

TABLE 3
EEC RELATIVE TOXICITY INDEX

| *Rating* | *Oral* $LD_{50}$ *(mg/kg), rat* |
|---|---|
| Highly toxic | 25 |
| Toxic | 25–200 |
| Harmful | 200–2 000 |

Testing for chronic toxicity is normally a long-term and expensive procedure, tests usually being carried out over complete life-spans. Currently such testing is restricted to chemicals under suspicion of being human carcinogens. Care is required in the interpretation of the results, however, since extrapolation from animals to man is not always simple. In more recent years a series of *in vitro* tests[22,23] (Ames for example) have been developed to provide a more rapid testing procedure. The value of the data from such tests is enhanced by obtaining the same positive or negative results from a series of different tests. Such battery testing is becoming more common.

Prior to 1970 legislation had been enacted in order to prevent exposure to hazardous substances in consumer products intended for household use. Food and cosmetic additives, pesticides, etc., are obvious examples, manufacturers of such commodities being required to supply evidence concerning the safety of such chemicals.

The availability of suitable test data and procedures facilitated the introduction of legislation, which included specified exposure values, or Threshold Limit Values. The TLV is that concentration to which nearly all persons may be exposed for an 8 h working day or 40 h week with no adverse effects. A more detailed discussion of TLVs, including definitions of the different categories, can be found in, for example, Heath and Safety Executive publication EH15/80.[24] The introduction of Health and Safety Acts in the 1970s considered the control of exposure to chemicals to protect not only the consumer but also the producer. Spivey[25] compares the implications of specific items of legislation as promulgated in Scandinavia (see Table 4).

The approaches taken by some different countries to the control of toxic substances have been reviewed in the literature, the USA,[26] the UK[27] and the European chemical industry[28] being specifically cited.

### *4.9.1 The Present Situation in the Rubber Industry*

Currently the suppliers of all rubber chemicals provide toxicological information, in the form of hazard data sheets, pertinent to their products. Handling procedures, special precautions, hazards and methods of dealing with spillage are some of the items recorded on such data sheets. Compendia of the toxicological properties[29,30] of rubber chemicals have been produced, for example *Toxicity and Safe Handling of Rubber Chemicals* (British Rubber Manufacturers Association Ltd) and *A Toxicological Survey of Chemicals Used in the Swedish Rubbery Industry* (National Board of Occupational Safety and Health, Sweden).

The importance of such information, or lack of it, was demonstrated when incidences of 'papilloma of the bladder' in the rubber industry were linked to the presence of β-naphthylamine in an antioxidant then in current usage. The detection of this carcinogen in the rubber factory atmosphere set the pattern for the development of the industry's hygiene to the present day, namely, changes in working procedures, legislation and analytical techniques. The decision to withdraw the antioxidants was a voluntary move by industry. In the UK the manufacture and use of β-naphthylamine is prohibited under the Carcinogenic Substances Regulations of 1967.

Hazards which are known to the industry may be adequately controlled. Toluene diisocyanate (TDI), a bulk material used in the

TABLE 4

COMPARISON OF LEGISLATION AND IMPLICATION

| *Legislation* | *Implication* |
|---|---|
| *Denmark* | |
| Threshold Limit Values, 1978 | Controls the maximum values of listed contaminants permissible in the workplace |
| Dangerous Substances and Preparations: Instructions Concerning Classification, Packing and Labelling, 1977 | Implements the EEC Regulations on the Packaging and Labelling of Dangerous Substances |
| *Norway* | |
| Threshold Limit Values, 1978 | Controls the maximum values of listed contaminants permissible in the workplace |
| Act Relating to Worker Protection and Working Environment, 1976 | Controls working conditions particularly with regard to toxic substances and defines responsibilities of suppliers of toxic substances, e.g. concerning labelling which must be in Norwegian |
| Product Control Act, 1978 | All chemicals supplied to industry or public must be registered |
| *Sweden* | |
| Threshold Limit Values, 1978 | Recommends values of air contaminants in the workplace enforceable by the Health and Environment Act, 1973 |
| Products Hazardous to Health and Environment Act, 1973 | Controls the manufacture, sale and handling of products hazardous to health and environment |
| General Product Regulation Act, 1978 | All products must be registered and eventually toxicological data will be required |

polyurethane industry, is known to cause respiratory problems and, in order to provide safe working, extremely well-ventilated conditions and specialised protective equipment are provided in accordance with the BRMA Code of Practice.[31] In this particular instance special monitoring equipment was developed and is now extensively utilised throughout the polyurethane industry.

More recently the toxicological hazards associated with the use of solvents, and benzene in particular, have been recognised. Epidemiological studies[32,33] have indicated a link between exposure to solvents and the development of leukaemia. Control of the situation was effected by a change to solvents considered to be less toxic and known to contain less than a certain proportion of benzene.

One approach adopted by the rubber industry to minimise exposure to chemicals has been that of local exhaust ventilation, the general aim being to maintain exposure levels below at least one-half of the TLV.

Ventilation is used in the weighing and mixing areas to prevent exposure to dusts. In curing areas it is essential to prevent volatile species from contaminating the workplace, and in general such fumes are exhausted outside the plant. From cost and efficiency considerations it is better to remove these contaminants from as near to the source as possible. There are several considerations which must be borne in mind: the extraction system should pull the fumes *away* from the operator and all ventilation systems should be adequately serviced. There are many ventilation manuals available and also one designed specifically to deal with situations to be met in the rubber industry.[34]

Despite all these control techniques, legislation and tests carried out in rubber factory atmospheres, epidemiological surveys still report excess deaths from, particularly, stomach and lung cancer.[35–38] Baxter and Werner[38] linked excess deaths from lung cancer with exposure to rubber fumes, with similar data being reported from the USA.[35] Cancer of the stomach has been associated, particularly, with occupations early in the production process where formulations were compounded.[38–41]

### *4.9.2 Future Developments*

Baxter and Warner postulated that the excessive lung cancer throughout the UK industry could be associated with the presence of a, so far, unidentified carcinogen in the workplace atmosphere. There are two different avenues of approach which are being used to investigate such a possibility.

The increasing sophistication and detection capabilities of instrumentation used in the analysis of the contaminants in a factory atmosphere are enabling the identification of chemical compounds hitherto not suspected of being present. The results from combined gas chromatograph/mass spectrometry investigations reveal a complexity of chemical compounds such that the application of an overall maximum level for fumes has been recommended by the BRMA.[40]

An alternative approach, which identifies potential on-site contaminants, has been developed at RAPRA.[41,42] In these experiments a compounded rubber is cured in a specially designed apparatus which facilitates the collection of the volatile species emitted. Care has been taken to ensure that the rubber experiences the same time and temperature profile as during normal processing. Using such equipment provides a means whereby repetitive and quantitative analysis of the process fumes can be carried out.

Investigations of this nature indicate that although the toxicological properties of the raw materials are well known, the chemical reactions involved in vulcanisation produce species which may bear no relationship chemically or toxicologically to the raw materials.

Over the past five years RAPRA has been extensively involved in the development of techniques to identify and quantify the species liberated during the process of vulcanisation. This work has led to an appreciation of the dramatic effects of cure temperature and the type of crosslinking agents used on both the yield and nature of the volatile emissions. The range of volatile species so far identified can be assessed from the examples given in Table 5.

TABLE 5
VOLATILE SPECIES

| RESIDUAL SPECIES IN POLYMERS | | DISTILLATION PRODUCTS | REACTION PRODUCTS |
|---|---|---|---|
| | | Residual polymerisation aids | Hydrogen sulphide |
| | | | Carbon disulphide |
| Butadiene | SBR | Extender oils | Amines |
| Styrene | SBR | Plasticisers | Mercaptans |
| Toluene | SBR | Other process aids | Isothiocyanates |
| Ethyl benzene | SBR | | Dimethyl nitrosamine |
| 4-Vinyl cyclohexene | SBR | | Nitrosomorpholine |
| etc. | | | etc. |
| Propene | EPDM | | |
| Ethylidene norbornene | EPDM | | |
| Dicyclopentadiene | EPDM | | |
| etc. | EPDM | | |

Possible routes to the production of dimethyl nitrosamine are the reaction of dimethylamine from TMTD with residual sodium nitrite[43] or NDPA[44] in the polymer. This interreaction of species is not unique. A reaction between carbon disulphide (from ZDMDC) and aniline (from

DPG) has been identified as the possible production route to significant levels of phenyl isothiocyanate in a plant atmosphere.

The few examples cited above serve to indicate the complexity of the chemical nature of the contaminants associated with rubber fumes. This much greater understanding of the nature of such fumes not only provides a new approach to the understanding of vulcanisation chemistry but may also supply a means whereby workplace fumes can be controlled. The elimination of potential routes to particular pollutants could lead to the prevention of pollution at source. Such an approach is currently being investigated by RAPRA.

Thus it can be seen that there is still much work to be done in identifying the pollutants in the rubber industry. This will not, however, be simply a case of cataloguing and investigation: the information can be used, hopefully, to eliminate major potential hazards.

### 4.10 Personal Hygiene

Exposure to chemical hazards, in particular, can be minimised by good personal hygiene. Good standards of hygiene are dependent on the provision of adequate facilities which should include:

(a) adequate washing, showering and changing facilities;
(b) an appreciation of potential sources of hazard;
(c) arrangements for the consumption of food and drinks in areas remote from chemical handling and storage;
(d) the prevention of smoking in areas in which chemicals are handled since there is the risk of both ingestion and also of creating a fire hazard;
(e) adequate protective clothing should be supplied to prevent skin contact, and it is also sensible to ensure that eyes are always protected.

Good standards of personal hygiene require co-operation on all sides, the provision of suitable facilities and reasons for the use of such equipment combined with a willingness to co-operate. It might be interesting to speculate on how many dust masks had been used to protect the operator's foreheads in the past!

### 4.11 Labelling and Identifying Hazardous Materials

The labelling and identification of hazardous material is a growing area and one in which there is as yet little international agreement. The discussions involve not only how a risk should be identified but also the nature and severity of the risk. This is particularly true for mixtures and dispersions of various compositions. It is, however, certain that clear and precise information detailing the particular hazards should be provided on all containers.

## 5 CONCLUSION

In this chapter an attempt has been made to discuss, almost at the general interest level, all those factors which might affect the healthy and safe operation of the rubber industry. A growing awareness of the nature of the newly identified hazards has been indicated, as has the increasing burden of legislation.

How will the industry respond to this situation? One approach may be the introduction of, and more emphasis being placed on, safer working practices, which will require information to be disseminated by publication, training and supervision. Implementation of such an approach will require the active co-operation of the regulatory authorities, trade federations, suppliers and all factory personnel.

New hazards will be identified and new techniques will be found to reduce their significance either by elimination or control. Such is the nature of progress.

## REFERENCES

1. Sax, N. I. (1979) *Dangerous Properties of Industrial Materials*, Van Nostrand Reinhold, New York.
2. Patty, F. A. (1980) *Industrial Hygiene and Toxicology*, Vols. I–III, Interscience, New York.
3. Handley, W. (ed.) (1977) *Industrial Safety Handbook*, McGraw-Hill, London.
4. Ragsdale, K. H., ibid., p. 319.
5. Illuminating Engineering Society (1973) *IES Code for Interior Lighting*, London.
6. British Electrical Development Association (1955) *Lighting in Industry*, London.
7. HMSO (1975) *Dust Explosions in Factories: Classified List of Dusts Tested for Explosibility*, SHW 830, London.
8. Palmer, K. N. (1973) *Dust Explosions and Fires*, Chapman & Hall, London.
9. Dept of the Environment (1972) *Code of Practice for Reducing the Exposure of Employed Persons to Noise*, HMSO, London.
10. US Dept of Labor (1969) Health and Safety Standards. *Fed. Register*, **34**(96), 29 May.

11. HARRIS, C. M. (1957) *Handbook of Noise Control*, McGraw-Hill, New York.
12. BERANEK, L. L. (1971) *Noise and Vibration Control*, McGraw-Hill, New York.
13. HOLMER, C. I. (1972) *Techniques of Plant Engineering and Maintenance*, **23**, 296.
14. BRITISH STANDARDS INSTITUTION (1975) *Code of Practice for Safeguarding of Machinery*, BS 5304, London.
15. INTERNATIONAL COMMISSION ON RADIOLOGICAL PROTECTION, publications from Pergamon Press, New York.
16. NATIONAL COUNCIL ON RADIATION PROTECTION AND MEASUREMENT (1971) *Basic Radiation Protection Criteria.* Report No. 39.
17. ACGIH (1980) *Threshold Limit Values for Chemical Substances and Physical Agents in the Workroom Environment*, Cincinnati, Ohio.
18. INGALLS, T. H. and ROBERTSON, J. MCD. (1980) *Arch. Env. Hlth.*, **35**, 181.
19. MCMICHAEL, A. J., SPIRTAS, R. and KUPPER, H. H. (1974) *J. Occup. Med.*, **16**, 459.
20. TROITSKAYA, N. A., VELICHKOVSKY, B. T., KOGAN, F. M. and KUZ'MINYKH, A. I. (1980) *Voprosy Onkologii*, **26**, 67.
21. QUALLINGTON, J. and DOLBEY, P. C. (1979) RAPRA Members Report No. 25, Shawbury.
22. AMES, B. N., MCCANN, J. and YAMASKI, E. (1977) *Mutat. Res.*, **31**, 347.
23. PURCHASE, I. F. H. (1980) *Br. J. Ind. Med.*, **37**, 1.
24. HEALTH AND SAFETY EXECUTIVE (1980) *Threshold Limit Values*, Guidance Note EH15/80, HMSO, London.
25. SPIVEY, A. M. (1979) *Kautsch. Gummi Kunstst.*, **32**, 766.
26. BINGHAM, E. (1980) *Ann. Occup. Hyg.*, **23**, 79.
27. PITTOM, L. A. (1980) *Ann. Occup. Hyg.*, **23**, 85.
28. SCHÖN, N. (1980) Proc. 21st Annual Meeting, International Institute of Synthetic Rubber Producers Inc., Mexico City.
29. BRMA HEALTH RESEARCH UNIT (1978) *The Toxicity and Safe Handling of Rubber Chemicals: A Code of Practice*, Birmingham.
30. NATIONAL BOARD OF OCCUPATIONAL SAFETY AND HEALTH (1977) Investigation Report 19, Stockholm.
31. BRMA HEALTH RESEARCH UNIT (1978) *Isocyanates in Industry: Operating and Medical Codes of Practice*, Birmingham.
32. MCMICHAEL, A. J., SPIRTAS, R., KUPPER, L. L. and GAMBLE, J. F. (1975) *J. Occup. Med.*, **17**, 234.
33. MCMICHAEL, A. J., ANDJELKOVICH, D. A. and TYROLER, H. A. (1975) *Ann. N.Y. Acad. Sci.*, **271**, 125.
34. RAPRA (1979) *Ventilation Handbook for the Rubber and Plastics Industries*, Shawbury.
35. FOX, A. J., LINDARS, D. C. and OWEN, R. (1974) *Br. J. Ind. Med.*, **31**, 140.
36. FOX, A. J. and COLLIER, P. F. (1976) *Br. J. Ind. Med.*, **33**, 249.
37. MONSON, R. R. and NAKANO, K. K. (1976) *Am. J. Epidemiol.*, **103**, 284.
38. BAXTER, P. J. and WERNER, J. B. (1980) *Mortality in the British Rubber Industries, 1967–76*, HMSO, London.
39. ANDJELKOVICH, D., TAULBEE, J., SYMONS, M. and WILLIAMS, T. (1977) *J. Occup. Med.*, **19**, 397.

40. McMichael, A. J., Spirtas, R. and Kupper, L. L. (1974) *J. Occup. Med.*, **16**, 458.
41. Willoughby, B. G. and Lawson, G. (1980) *Rubb. Chem. Technol.*, **54**, 311.
42. Willoughby, B. G. and Lawson, G. (1979) Paper 29, ACS Rubber Division Meeting, Cleveland, Ohio.
43. McMichael, A. J., Spirtas, R., Gamble, J. F. and Tonsey, P. M. (1976) *J. Occup. Med.*, **18**, 178.
44. Ashness, K. G., Wetton, R. E. and Willoughby, B. G. (1980) *Kautsch. Gummi Kunstst.*, **33**, 1007.

*Chapter 5*

# NATURAL AND SYNTHETIC RUBBER IN SPORTS GOODS

R. C. HAINES

*Dunlop Sports Company Limited, Wakefield, Yorkshire, UK*

*SUMMARY*

*Balls for use in golf, tennis and squash use significant amounts of natural and synthetic rubber which is now calculated to total in excess of 10 000 tonnes/year. Owing to the traditional aspects of games, changes in ball composition and construction are strictly limited both by specification and player acceptability and so developments have occurred only slowly. However, of recent years new materials and manufacturing techniques have been increasingly applied to improve ball performance and durability and also to limit the inevitable increase in manufacturing cost.*

## 1 INTRODUCTION

The special properties of rubber have long made it attractive for use in sports and games, and its first reported use in the form of a gum-rubber ball by natives of South America is doubtless accurate. The well-known property of rubber to deform to a high degree and to snap back to its original size and shape with minimum energy loss leads to its ability to bounce and so produce a plaything of fundamental attraction to man. It is this quality which is the basis of its use in balls for three of the most popular individual (as distinct from team) games of tennis, golf and squash.

Other properties of rubber, such as flexibility, low gas permeation, high friction, etc., have determined its use in a wide range of sports accessories as well as for bladders for footballs and the like: sports shoes, wet suits and flippers for activities on and under water, table tennis bat surfaces, grips for golf clubs—the list is endless.

Special technology has been perfected over the years in the manufacture of balls for tennis, golf and squash and because of the high number of these balls made each year, this chapter will be devoted to these products. It is estimated that 300 million golf balls, 180 million tennis balls and 6 million squash balls were made throughout the world in 1980, and although squash ball manufacture is considerably smaller than in the case of the other two balls, the game is expanding rapidly, particularly in Europe.

It is important that, in considering the manufacturing and development aspects of games equipment, the traditional aspects of games should be borne in mind because this very much restricts aspects of product development. Development usually means change, and this conflicts with the historical and traditional aspects which are jealously guarded by controlling authorities who exercise their powers to ensure uniformity of equipment and procedures not only from a geographical standpoint, but also over a considerable time-span. This means that product and manufacturing improvements have to take place within the strict confines of equipment specifications, the rigour of which varies from one authority to another.

Reference must also be made to factors associated with equipment qualities which are intangible and difficult to define in purely scientific terms and yet are of extreme importance from the player's viewpoint for a full and satisfying appreciation of the game he plays. For instance, the 'click' of a golf ball on contact with a clubhead, the 'response' of a ball on the racket strings, the tactile nature of a tennis ball surface are all cases in point which defy specification. Such qualities can determine whether an otherwise satisfactory product is successful commercially, and so manufacturers spend a considerable amount of energy in developing their own particular expertise and 'know-how'.

It should be noted that plastic materials have made and are making considerable inroads into equipment for sports and games and in any case the boundary between rubbers and plastics is becoming more obscure. Because this chapter is concerned with rubber and rubber composites, plastic materials will only be referred to where they are an important constituent in what is basically a rubber product.

## 2 TENNIS BALLS

### 2.1 Historical

The game of tennis (more properly know as lawn tennis) was developed in the 1870s from the game of royal or real tennis which has been played for centuries. While both games are similar, in that they are both based on one or a pair of players opposing each other from different sides of a net hitting a ball back and forth, the type and size of court, method of scoring, etc., are very different. Real tennis balls have been traditionally made from a spherical stitched envelope of leather or cloth stuffed with rags, horsehair or similar material, while tennis balls have always been based on rubber.

Originally tennis balls were made solely of rubber, but the wearing and playing properties of the balls were improved by covering them with flannel stitched around the rubber 'core'. The ball was quickly developed by making the core hollow and pressurising it with gas. Originally, core manufacture was based on the 'clover-leaf' principle whereby uncured rubber sheet was stamped into a shape resembling a three-leaf clover and this was assembled into a roughly spherical shape by machinery adapted for the purpose. Chemicals generating pressurising gas were added prior to closing the assembly and these were activated on moulding the core to a spherical shape in heated cavities. This process was used for many years until the precision of the game demanded a higher degree of uniformity (particularly relating to wall thickness) than could be obtained with the clover-leaf method. Now it is usual to compression mould two separate 'half-shells' which are assembled together to produce a 'core'. The original flannel cloth was replaced by special 'Melton' cloth made specifically for the purpose and the stitching has been replaced by a vulcanised rubber seam.

### 2.2 Tennis Ball Specifications

The ball specification is set by the International Tennis Federation and it covers ball weight, size, rebound and 'compression', tested at 20 °C and 60% relative humidity as follows:

| | |
|---|---|
| Weight | between 2 oz (56·7 g) and $2\frac{1}{16}$ oz (58·5 g) |
| Size (go/no go ring gauge) | between 2·575 in (6·54 cm) and 2·700 in (6·86 cm) |

| | |
|---|---|
| Rebound (from 100 in on to concrete) | between 53 in (135 cm) and 58 in (147 cm) |
| Compression: | |
| Forward deformation | between 0·220 in (0·56 cm) and 0·290 in (0·74 cm) |
| Return deformation | between 0·350 in (0·89 cm) and 0·425 in (1·08 cm) |

The compression test procedure requires the use of a modified 'Stevens' machine[1] which is designed to measure the deformation of a ball under a load of 18 lbf (8·165 kgf) both when the load is initially applied (forward deformation) and on recovery (return deformation) after the ball has been subject to deformation of 1 in (2·54 cm). For a perfectly elastic ball the forward and return readings would be the same and so the difference between the readings gives a measure of the hysteresis loss of the ball. Prior to compression testing it is stipulated that the ball must be deflected through 1 in (2·54 cm) on each of three diameters at right angles. This is designed to take any 'set' out of the ball which can occur after prolonged storage.

### 2.3 Modern Manufacture

The manufacturing operations consist basically of the following:

1. Half-shell moulding.
2. Assembly of half-shells to produce a core.
3. Preparation of Melton.
4. Covering of core with Melton.
5. Cover vulcanising.
6. Finishing operation.
7. Testing and marking.

#### *2.3.1 Half-Shell Manufacture*

This is normally carried out by a simple compression-moulding operation using accurately prepared blanks, for instance, as prepared by a Barwell Pre-Former. For speed of operation it is normal to load the press from preloaded jigs which enable a large number of cavities to be loaded simultaneously. Again, to aid handling, it is normal for the flash from the individual cavities to join together so that the press can be quickly stripped, and the material wastage in terms of flash is more than compensated for by operational speed. A blanking-out operation removes flash from individual half-shells.

### *2.3.2 Core Manufacture*

The edges of the half-shells are lightly abraded where they are to be jointed and vulcanising rubber solution is applied. The process then differs according to whether chemical or compressed-air methods are to be used for inflation.

The chemical method is older and while it has the advantage that the gas pressure (and hence core properties) can be accurately controlled from the quantity of chemicals used, and the fact that inflation occurs after the core has been assembled, it suffers from the disadvantage that under certain conditions of storage, products of the reaction can crystallise out and cause 'rattlers'. The moulding process is by simple compression moulding using mating hemispherical cavities. The inflation chemicals are normally sodium nitrite and ammonium chloride which produce nitrogen during the moulding process.

The compressed-air inflation method uses a complex clam-shell press arrangement in which the individual solutioned half-shells are placed separately in opposite mating hemispherical cavities of the press so that, on closing the press, the half-shells opposite each other are brought together to form a core (Fig. 1). However, the closing operation takes place in stages as follows:

(a) The press closes until the edges of the half-shells are not quite touching.
(b) In this position the internal platen area is isolated from the atmosphere by a rubber sealing ring around the periphery of the platens.
(c) Compressed air is introduced at the required pressure to the platen area.
(d) The press platens are then brought together, so trapping the compressed air between the half-shells and forming a pressurised core.
(e) The platens are then heated to vulcanise the rubber cement and then cooled prior to core extraction.

The mould and press arrangement for compressed-air inflation involves a considerably larger mass of metal than for chemical inflation, and process times tend to be longer. Also, great care is necessary to ensure that the cores are equally pressurised over the whole platen area, and so good equipment maintenance is necessary. The compressed-air inflation method is applicable to other gases which might be attractive for specific purposes (see below).

FIG. 1. Cores being removed from an air-inflation press. (Courtesy: Dunlop Sports Co. Ltd)

*2.3.3 Formulations*

Core formulations for pressurised balls are normally based on natural rubber containing a high loading of fine-particle filler for low gas permeability. A formulation (parts by weight) is as follows:

| | |
|---|---|
| Natural rubber | 100 |
| GPF black | 30 |
| Clay | 32 |

| | |
|---|---|
| Zinc oxide | 9 |
| Sulphur | 3·5 |
| DPG | 2 |
| HBS | 1 |
| Cure: Half-shell | $2\frac{1}{2}$ min at 150 °C |
| Core (chemical inflation) | 4 min at 150 °C |
| (air inflation) | 12 min at 150 °C |

Cores are normally pressurised to around 12 lb/in$^2$ (0·84 kg/cm$^2$), but owing to the fact that rubber compounds are permeable to gas, the pressure is gradually lost and balls lose rebound and compression so that after some 4 months they no longer meet specification. For longer storage periods it is normal to put the balls in pressurised cans, and almost all balls sold in the USA are so packed. However, so-called pressureless (non-pressurised) balls have been developed and have sold in some markets for many years. The basic technology of making a pressureless ball is that of producing highly resilient, high-modulus compounds of lower than normal density such that they can be used to produce cores with walls of greater than normal thickness which can provide unaided the necessary compression and rebound properties. Otherwise manufacturing procedure is the same as for pressurised balls.

Compounding methods for pressureless balls described in patent literature are varied and involve the use of high styrene resin,[2] cellulosic filler,[3] aminoplast resin,[4] copolymer of an olefin and an unsaturated monocarboxylic acid copolymer.[5] The various formulations attempt to produce appropriate qualities of ball performance and player satisfaction as well as fulfilling the necessary specification requirements. In practice, pressureless balls have had only limited commercial success which is reflected by the fact that well under 10% of balls currently produced in the world are of the pressureless variety.

### *2.3.4 Core Manufacturing Developments*

Since the days of the clover-leaf core, attempts have been made to make a precision core by avoiding the separate operations of half-shell and core moulding. Methods of producing uncured half-shells which can be assembled into pressurised cores in only one vulcanising operation are described in patent literature.[6–8] The main problem is that of producing a strong joint between the halves of the core while producing minimum local distortion from the inevitable pressure which must be applied to effect jointing of the unvulcanised material.

### *2.3.5 Melton Cloth*

Special Melton cloths have been developed for tennis balls. Best-quality tennis balls use Melton with a high wool content (50–65%) for three primary reasons:

(a) to enable the nap of the cloth to be satisfactorily consolidated by fulling (or felting);
(b) to allow the easy removal of moulding lines after the cover moulding operation;
(c) to give low soiling properties in play.

For optimum wearing properties, the balance of the fibre content is made up by nylon. No other synthetic fibre has challenged nylon in this respect.

The Melton cloths are produced by the long-established processes of weaving, raising and fulling as follows:

(a) *Weaving*. The basic cloth is produced by weaving blended wool/nylon weft yarns into a light cotton warp where the warp weight normally constitutes only about 15% of the final cloth weight. A 'Sateen' weaving pattern is used in which the weft lies predominantly on one side of the warp and it is this that becomes the wearing surface of the cloth. Cloth weight for a best quality ball is about 700 $g/m^2$.
(b) *Raising*. The 'face' of the cloth is then subject to a 'raising' operation in which the fibres of the weft are carefully teasled to produce a fluffy texture without destroying the basic integrity of the yarns.
(c) *Fulling*. The raised cloth is then worked in water containing soap so that the raised wool fibres become felted together to produce a consolidated surface which ultimately becomes the wearing surface of the ball.

Considerable skill is exercised by the cloth manufacturers, particularly in the raising and fulling processes, so that, in play, the cloth surface generates and maintains as long as possible the appropriate surface texture which is required for desirable ball flight qualities and end-wear. In achieving this objective, account must be taken of the widely differing surfaces on which tennis is now played, ranging from grass on the one hand through clay and shale and profiled plastic to highly abrasive 'all-weather' surfaces on the other.

*2.3.6 Needle Cloth*

While Melton cloths are used predominantly for top-quality balls, so called 'needle' cloths have been and are being used, but mainly for lower grades. This cloth is made by a process of 'needle punching' which produces a consolidated fabric with the general appearance of a Melton directly from staple fibres by avoiding the stages of yarn preparation, weaving, raising and fulling. In the needle process, a mat of randomly orientated fibres (sometimes assembled on a light 'scrim' cloth) is passed through a needle loom which is made up of a multitude of barbed needles which are made to reciprocate up and down. The barbs on the needles point downwards and as they pass into the mat of fibres, individual fibres are picked up by the barbs and are pushed downwards where they are left as the needle withdraws. The action of hundreds of needles acting in this way gradually consolidates the fibres into a continuous interlocking structure which depends for its integrity on inter-fibre friction.

While such needle cloths are cheaper to produce than Melton, they have the disadvantage that they tend to be less uniform in thickness and tend to lose surface nap more quickly. Because fabric consolidation does not depend on the felting properties of wool, high synthetic fibre contents can be used, but mould line removal after ball moulding then becomes the controlling factor.

*2.3.7 Cloth Preparation*

The cloth is normally coated on its reverse side with a vulcanising cement by means of one or more spreading operations. 'Dumbbell'-shaped blanks are then cut from several layers of cloth at once by means of an automatically indexing stamping press (Fig. 2) with the axis of the blank being aligned at approximately 45° to the warp direction so that the dumbbell can stretch axially during the covering operation.

The edges of the blanks are next coated with a vulcanising cement which subsequently forms the seams of the ball. This is normally done by first putting the blanks into a stack, compressing the blanks tightly together into a jig and then immersing the assembly in a vat of vulcanising rubber solution. The immersion is normally done by an automatic process in which jigged stacks are taken by overhead conveyors to one or more dipping and draining stations.

*2.3.8 Ball Covering*

Prior to the covering operation, the cores are subject to surface abrasion

FIG. 2. Melton blanks being stamped from layers of Melton by an automatically indexing cutter. (Courtesy: Dunlop Sports Co. Ltd)

and are then coated with rubber solution in a barrelling operation in which cores together with a measured volume of rubber solution are tumbled together, so coating each core with a uniform layer of solution.

The covering operation is carried out by hand (Fig. 3) or machine using two dumbbell blanks for each core. The difficulty of the operation can be appreciated in view of the fact that the cloth can stretch, the core is deformable and the tackiness of the rubber cement on core and cloth produces adhesion on contact.

After covering, the balls are subject to a cover moulding operation to vulcanise cloth to core and the joint between the blanks. This is carried out at a temperature which avoids the heat discoloration of the cloth and is followed by a cooling process as in the case of all inflated products.

### *2.3.9 Finishing Operations*

The covered ball moulding operation compresses the nap of the cloth and the split line of the mould produces a crease; both effects are removed by a steaming operation in which the balls are tumbled in wet

FIG. 3. Core being covered by hand. (Courtesy: Dunlop Sports Co. Ltd)

steam. After stabilising, the balls are tested for compression and printed by a hot transfer method.

### 2.4 Future Developments

While developments involving both core and cloth have occurred over the years, the inflation medium has received little attention following the development of the compressed-air inflation method. However, it is known that certain gases, because of their molecular complexity, permeate through rubber at a significantly lower rate than air or nitrogen and such gases could have application in tennis ball cores. Such a gas is sulphur hexafluoride,[9] which has become cheaply available over the last few years because of its application in certain electrical equipment. Pressurised cores with a storage life increased by the order of a factor of four are promised by this development, but technical difficulties related to the uniform pressurisation of cores with this gas need to be overcome.

## 3 SQUASH BALLS

### 3.1 Historical

The game of squash rackets (squash for short) is thought to have been developed at Harrow School 100 or so years ago. Boys awaiting their turn on a rackets court are reported to have played with a rackets ball in an adjacent room or corridor, but having broken windows with the hard ball, they resorted to using a rubber ball of similar size. The modified game (no doubt called squash rackets in view of the 'squashier' nature of the ball) spread to other public schools and then to the Services. The Squash Rackets Association was formed in England in 1929 where the rules were standardised, and an International Squash Rackets Federation was formed in 1967. The game has developed differently in the USA where, amongst other differences, larger, harder balls are specified.

### 3.2 Squash Ball Specification

The ball specification set by the International Squash Rackets Federation is as follows:

| | |
|---|---|
| Weight | Between 23·3 g and 24·6 g |
| Size (go/no go gauge) | Between 39·5 mm and 41·5 mm |
| Rebound (from 100 in on to concrete at 68 °F) | Between 18 in (45·72 cm) and 24 in (60·96 cm) |
| Compression | Deformation under a load of 2·4 kgf of between 3 and 7 mm |

The rebound specification is in fact subject to subdivision to accommodate balls of different 'speeds' identified by different colour markings applied to the ball, as indicated below:

| | |
|---|---|
| 18 in (45·72 cm) to 20 in (50·8 cm) | Yellow mark |
| 20 in (50·8 cm) to 22 in (55·88 cm) | White |
| 22 in (55·88 cm) to 24 in (60·96 cm) | Red |
| over 24 in (60·96 cm) | Blue |

It is important to note that the appropriate playing qualities of a squash ball cannot be completely and adequately described by the specification, and the skill in producing a good squash ball lies in the compound formulation which provides (in addition to the specified requirements) satisfactory rebound characteristics at a range of impact speeds and a range of playing temperatures. In addition, appropriate surface friction is necessary to produce acceptable rebound angles, and minimum change in this surface friction over a reasonable playing life is desirable. The necessity for good durability is particularly important in view of the extremely high deformation to which the ball is subject in play (Fig. 4). This has particular significance in terms of the strength and durability of the joint between the half-shells from which the ball is made.

Squash balls have traditionally been made of carbon black loaded compounds and the fact that these balls mark the walls of the squash court (and so necessitate wall cleaning at frequent intervals) was accepted. With the recent rapid expansion of squash and the building of new courts this has become unacceptable, and so-called 'non-marking' balls are now produced based on non-black compounds. Generally, however, the black-type balls are preferred for championship play because of their particularly subtle playing qualities and they are the standard against which the other balls are assessed. Attempts to produce non-marking balls with playing characteristics indistinguishable from those of championship balls have not been completely successful, but it is the ultimate objective of the squash ball compounder.

### 3.3 Squash Ball Manufacture

Squash ball formulations are jealously guarded by manufacturers who develop their own manufacturing expertise and know-how. The ball is made in a similar manner to that of a tennis ball core with the separate stages of half-shell manufacture, ball assembly and final vulcanisation. Appropriate surface abrasion is produced by high-speed rotation in an abrasive barrel followed by washing. Surface marking is applied by a pad printer.

FIG. 4. High-speed photograph of a squash ball at the instant of impact on a racket. (Courtesy: Dunlop Sports Co. Ltd)

## 4 GOLF BALLS

### 4.1 Historical

Before 1848 golf balls were made by stuffing feathers under high pressure into a stitched leather container which was subsequently hammered into a spherical shape. In that year it was found that gutta-percha could be easily softened in hot water and fashioned to make a golf ball which generally performed better than the 'feathery'. In addition it was found that the inevitable surface cuts and indentations produced on the gutta-percha ball by the clubhead actually improved the flight of the ball, and so the science of golf ball aerodynamics was born.

The feathery existed for a few years after the development of the 'gutty', but the gutty itself was made obsolete around 1900 by the development in the USA of the 'Haskell', a ball with a core made from windings of stretched rubber thread. This ball has remained unchanged in concept until today, but it is now under pressure from modern non-wound balls based on the latest polymer technology, this time from non-wound cores with a separate cover—so-called 'two-piece' balls.

The significant detail developments of the wound ball over the years since its inception include the development of a liquid centre for the windings, improved rubber thread for greater resilience, fast and efficient winding machines, vulcanisable gutta-percha/balata cover compounds, synthetic balata (*trans*-polyisoprene), 'ionomer' cover materials, better paint and improved aerodynamics.

## 4.2 Golf Ball Specifications

Two separate authorities control the game of golf throughout the world in terms of the equipment used: the United States Golf Association (USGA) in the USA and Canada, and the Royal and Ancient Golf Club of St Andrews (R & A) elsewhere. While ball specifications are identical regarding weight, the two authorities specify balls of different size as indicated below:

| | | |
|---|---|---|
| Ball weight: | R & A, USGA | not greater than 1·62 oz (45·93 g) |
| Ball size: | R & A | not less than 1·62 in (41·15 mm) |
| | USGA | not less than 1·68 in (42·67 mm) |

Both authorities also specify a maximum ball resilience in terms of the speed of projection of the ball from a specially designed apparatus,[10] in which the ball is struck by a hammerhead moving at a controlled speed. On this machine the ball is restricted to a maximum velocity of 255 ft/s (77·2 m/s) at a temperature of 75 °F (24 °C). The maximum velocity rule was introduced in 1956 by the USGA but was adopted only in 1976 by the R & A, when it recognised that potential technological development might produce balls that could be hit further and so diminish the challenge of established golf courses.

It is interesting to note that only maximum ball weight and minimum ball size are referred to in the weight and size specification; this arises from the fact that, other things being equal, the smaller and heavier the ball within limits, the further it can be hit. Manufacturers are therefore encouraged to produce balls to the maximum allowed weight and minimum allowed size for ultimate performance.

A recent additional specification[11] has been introduced by the USGA and the R & A, and this requires that the ball shall be 'spherically symmetrical'. This is intended to prevent manufacturers developing balls which might produce directional flight qualities related to certain asymmetry of weight, surface pattern, etc., in relation to the direction of ball rotation and projection.

### 4.3 Golf Ball Compression

The compression of a golf ball (i.e. its resistance to deformation under static load) is not specified by the R & A or the USGA and it is left to manufacturers to produce balls to meet players' requirements. The most widely accepted format for expressing compression is referred to an arbitrary scale related to a ball's deflection under a load of 200 lb weight. A deflection of 0·1 in is counted as 100 and a deflection of 0·2 in and greater as zero. On this scale, balls favoured by professional golfers are in the 90–100 range. In general, balls of harder compression have a higher resilience and, other things being equal, manufacturers attempt to produce a large percentage of balls in the hard compression range. In the case of wound balls, this is limited by the tension at which the thread can be wound.

### 4.4 Wound Golf Balls

Wound golf balls are made from tightly stretched rubber thread wound on to a 'centre' which is necessary to provide weight, initiate the winding operation and to allow the free distortion of the core on impact so that high resilience can be generated. Until recently, the best golf balls incorporated 'liquid' centres almost exclusively because such centres can distort readily on impact and cause minimum energy loss. The ball is completed by moulding on a protective cover which incorporates the surface pattern (Fig. 5).

#### *4.4.1 Golf Ball Centres*

The original liquid centres consisted of a variety of fluids contained in a latex bag tied at its neck and bound with wide rubber tape prior to the

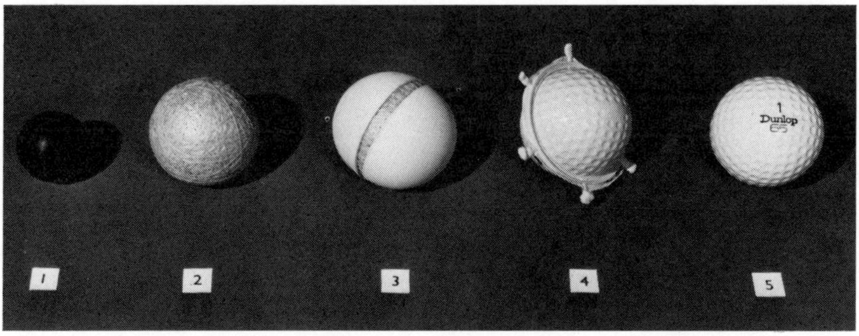


FIG. 5. The components of a wound golf ball: 1, centre; 2, wound core; 3 moulding assembly; 4, moulded ball; 5, finished ball. (Courtesy: Dunlop Sports Co. Ltd)

winding operation. A modern method for liquid centre manufacture is as follows:

(a) A suspension or 'paste' of barytes and bentonite clay is produced in water and glycerine and this is dispensed into a series of hemispherical cavities in a rubber mould. Two such moulds are located together so that spheres of paste are formed and the moulds are suitably jigged and passed through a freezer.
(b) On leaving the freezer, the frozen spheres are demoulded and encapsulated in unvulcanised rubber by a die-forming operation.
(c) The encapsulated spheres are allowed to thaw and are then subject to a compression-moulding process to vulcanise the covering.
(d) Before winding can commence, the centres are rotated in a barrel at a low temperature to freeze the paste again and to make them perfectly spherical for the winding operation.

With the advent of *cis*-polybutadiene rubber, highly resilient solid centres could be made by a simple compression-moulding process and such centres have been progressively replacing liquid centres since the mid-1960s. Because of their high resilience, such centres cause little energy loss in the ball and in fact their resilience is additive to that of the windings, so balls of increased resilience can be produced.

The solid centre must again be frozen prior to winding so that its spherical shape is preserved, but with certain winding machines this is unnecessary. A typical formulation (parts by weight) for a 1·62 in ball solid centre is:

| | |
|---|---|
| *cis*-Polybutadiene | 100 |
| Barytes | 160 |
| Dicumyl peroxide (40% active) | 5 |
| Cure: 15 min at 150 °C | |

*4.4.2 Thread and Winding*

Golf ball thread can be made from suitably compounded latex,[12] but the greatest volume of thread is made by the established dry rubber processes of mill mixing, calendering, drum curing and multi-knife slitting (Fig. 6). Conventionally, the thread is made to dimensions of 0·020 × 0·060 in (0·5 × 1·5 mm) or similar. The thread is compounded to give an elongation of about 1000% with high tensile strength and good chafe resistance so that it can be readily handled on the winding machine. A typical thread formulation (parts by weight) is as follows:

| | |
|---|---|
| Natural rubber | 65 |
| *cis*-Polyisoprene | 35 |
| Aldehyde/amine | 0·5 |
| Zinc oxide | 3 |
| 4,4′-Dithiodimorpholine | 1 |
| Sulphur | 1 |
| Cure: 180 min at 135 °C | |

A modern winding machine[13] is shown in Fig. 7. It consists basically of two driven rollers on which the core sits. The rollers not only rotate but they oscillate axially in antiphase so that the core is oscillated about a vertical axis as well as being rotated about a horizontal one. Two idler rollers press down on to the core under the action of a pneumatic cylinder which is programmed to release pressure on the core periodically. Thread is fed to the core from a spool or box and it is tensioned by passing around a rubber-covered roller controlled by an electromagnetic brake, the drag of which can be adjusted to stretch the thread to an appropriate degree.

The action of the machine is to rotate the core in a random manner under the action of the driven rollers, the idler rollers and the tension exerted by the thread. In practice, the thread is always wound on to the smallest diameter at any instant and the core can reorientate itself whenever the pressure exerted by the idler rollers is reduced. A micro-switch stops the machine when the core grows to the correct size.

FIG. 6. Slitting golf ball thread. (Courtesy: Dunlop Sports Co. Ltd)

*4.4.3 Cover Moulding*

Until the 1960s, golf balls were almost exclusively covered with gutta-percha/balata compositions because they could be moulded at temperatures (85–96°C) which did not degrade the highly stretched thread. After 1960, synthetic balata (*trans*-polyisoprene) became available and this was and is used in blends with balata (gutta-percha is no longer available). At about the same time a new thermoplastic copolymer was developed[14] and was given the generic name 'ionomer'. It was characterised '... by having ionic crosslinks in the solid state which are removable on fusion and restored on reversion to the solid state'. The copolymer named 'Surlyn' (DuPont registered trademark) was based on ethylene and a carboxylic acid and for the first time it offered a man-made thermoplastic cover material which could be processed at reasonable temperatures to produce a good, tough ball cover.[15]

The moulding processes for balata- and ionomer-covered balls are distinctly different in that, because of the higher processing temperatures of ionomer (160°C), special moulding procedures are adopted (see below).

FIG. 7. Golf ball winding machine. (Courtesy: Dunlop Sports Co. Ltd)

(*a*) *Balata cover.* Half-shells are formed by compression moulding from suitably compounded balata/*trans*-polyisoprene material and these are assembled on to the wound core. The assembly is compression moulded at 90 °C followed by a cooling process involving the use of refrigerated water to aid extraction. The equator spue-line is then carefully ground off and the ball cover is subjected to a vulcanising operation. Commonly cover vulcanisation is carried out by exposing the balls to carbon disulphide vapour for several hours. The carbon disulphide reacts with a secondary amine and zinc oxide in the cover compound to produce zinc diethyldithiocarbamate and vulcanisation is completed by standing the balls in ovens for 48 h at 32 °C.

(*b*) *Ionomer cover.* Several different grades of Surlyn are available and cover compositions having desirable characteristics have been evolved by blending one or more different types together[16] or by blending with other thermoplastic materials.[17] The half-shells are produced by injection moulding and the opportunity can be taken to arrange for half-shells to be linked together for ease of handling.

Because of the higher moulding temperatures necessary with ionomers, a special moulding procedure is necessary to avoid thread degradation in the wound core. The core/cover assemblies are put into a hot mould, the mould is closed rapidly and cooling applied immediately on closure. Special mould coatings are necessary for good release properties (Fig. 8). Owing to the thermolabile crosslinks, no subsequent process of crosslinking is necessary as in the case of balata compounds.

#### 4.4.4 *Finishing*

Polyurethane paints, developed specifically for golf balls, have been available since the early 1960s and these have been subject to continuous improvement. For satisfactory paint adhesion a chlorination process is necessary in the case of balata-covered balls, but for Surlyn balls specially developed treatments have been devised involving abrasive treatment and/or surface oxidation combined with the use of special paint primers. Ball marking is normally carried out by foil printing, but pad printing and transfer methods are also used. The surface is protected by a final polyurethane lacquer coat.

### 4.5 Non-Wound Golf Balls

#### 4.5.1 *Solid Balls*

Since the demise of the solid gutty ball around 1900, attempts have been made to produce improved non-wound balls, but such attempts were generally unsuccessful until developments in polymer technology provided the appropriate materials. The problem of producing a good non-wound ball is one of producing a hard, highly resilient material which is tough enough to withstand the repeated impacts demanded of a golf ball. Such compositions were identified by James Bartsch who in 1963 filed claims for a homogeneous moulded golf ball '... predominantly composed of a filled elastomer highly crosslinked into a

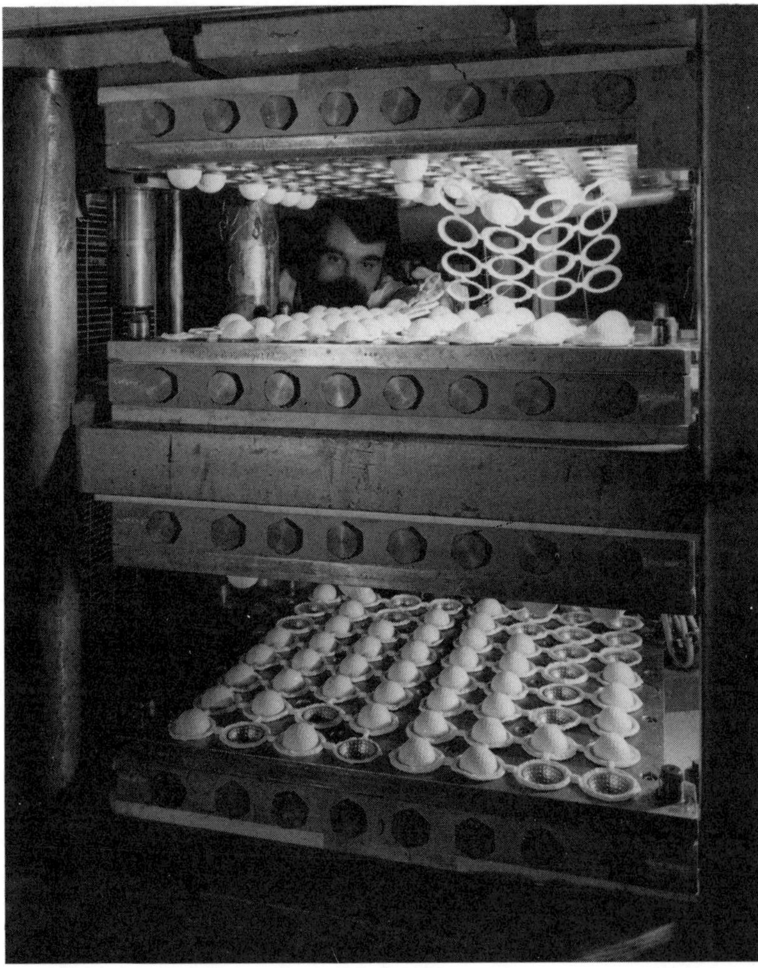

FIG. 8. Removing golf balls after moulding. (Courtesy: Dunlop Sports Co. Ltd)

three-dimension network with long flexible crosslinks formed from a polymerised crosslinking monomer, said elastomer being one capable of being so crosslinked by said crosslinking monomer'.[18] The most appropriate monomers are esters of unsaturated acids and those preferred are acrylates and methacrylates. These are used in conjunction with *cis*-polybutadiene, and a typical formulation (parts by weight) for a

1·68 in ball is shown below:

| | |
|---|---|
| *cis*-Polybutadiene | 100 |
| Butyleneglycol dimethacrylate | 62·5 |
| Silica | 62·5 |
| Dicumyl peroxide | 3·13 |
| Cure: 10 min at 150 °C | |

While Bartsch-type balls were superior to anything made previously, they did not have the resilience of a wound ball and were of the order of 10% short in distance. They have found application in driving ranges where their durability rather than distance is attractive.

*4.5.2 Two-Piece Balls*

The work of James Bartsch spurred on further developments in non-wound ball technology and this resulted in the advent of balls with non-wound cores (i.e. solid cores that required the protection of a separate cover because of their limited toughness), termed two-piece balls to define them from simple solid balls (one-piece) and conventional wound balls (three-piece).

The technology of two-piece balls is again based on the use of a cross-linking monomer, but in this case the preferred monomers are metal salts of polymerisable organic acids. Materials that have been found to be particularly suitable are zinc diacrylate and zinc dimethacrylate. They may be formed either *in situ* (i.e. during the actual compounding operation) or in a separate operation in which they are partially polymerised and are then added in the normal compounding procedure. The *in situ* process tends to give superior ball properties, but the prior partial polymerisation leads to a more convenient compounding procedure.

Generally, the 'metal-containing polymerisable monomers' produce cores of higher resilience than Bartsch-type monomers but they tend to be more brittle. For this reason they need the extra protection afforded by a separate 'cover' and so the two-piece format was devised (Fig. 9). A core formulation (parts by weight) suitable for a 1·68 in ball is given below:

| | |
|---|---|
| *cis*-Polybutadiene | 100 |
| Zinc diacrylate | 35 |
| Zinc oxide | 18 |
| Dicumyl peroxide (40% active) | 5 |
| Cure: 20 min at 160 °C | |

FIG. 9. Sections of two-component golf balls. (Courtesy: Dunlop Sports Co. Ltd)

The availability of ionomers allowed the possibility of directly injecting the cover on to the core—a process which is basically untenable for a wound core because of the high melt temperatures necessary. The direct cover injection method requires the accurate location of the core in the centre of the mould, and a method[19] involving the use of retracting supporting pins in the ball mould has been perfected. Typically, this is done with a vertical clamp insert moulding machine as shown in Fig. 10. Four cavities are installed in the mould which shuttles in and out of the injection position for ease of core insertion and ball extraction. Surface witnesses of the support pins and injection points are removed by a centreless grinding process, otherwise the finishing procedure is as for an ionomer-covered wound ball.

The performance of a two-piece ball is characterised by a lower flight trajectory than that of a wound ball, and this arises because of the higher moment of inertia of the ball around its centre point compared with a wound ball. Less spin is imparted by a lofted club, resulting in a flatter trajectory and less acute angle of impact which gives the two-piece ball greater run on striking the ground. Generally the characteristics of the ball are more suited to the longer handicap golfer, the better golfer

FIG. 10. Vertical clamp injection-moulding machine for direct cover injection of two-piece balls. (Courtesy: Dunlop Sports Co. Ltd)

preferring the stopping effect of a wound ball (arising from its greater spin) on and around the green. However, two-piece balls are now of considerable commercial importance and a significant percentage of balls made in the USA are of this type.

### 4.6 Golf Ball Moulds

The significant improvement in ball performance in recent years is due not only to material and process improvements but also to the accuracy with which golf ball moulds can be made. This has led to improvements in the aerodynamics of balls as determined by the surface pattern where, for instance, variations of 0·001 in in dimple depth on the ball are significant. Various procedures have been adopted for manufacturing moulds, but the most modern techniques are based on hobbing[20] or otherwise pressing[21] the desired surface profile into the mould surface so that a large number of moulds can be accurately reproduced.

FIG. 11. Golf ball distortion at the moment of impact. (Courtesy: Dunlop Sports Co. Ltd)

### 4.7 Golf Ball Aerodynamics

The complexities of golf ball aerodynamics are outside the scope of this chapter, but reference must be made to the subject in terms of golf ball surface patterns. The objective of the surface pattern is to produce aerodynamic 'lift' to the ball as it is caused to spin by the oblique contact of the lofted club head. In general terms, aerodynamic lift and drag (i.e. forces perpendicular to and along the direction of flight) increase together owing to the rotation of the 'boundary' layer of air immediately adjacent to the ball surface which is influenced by the surface pattern.

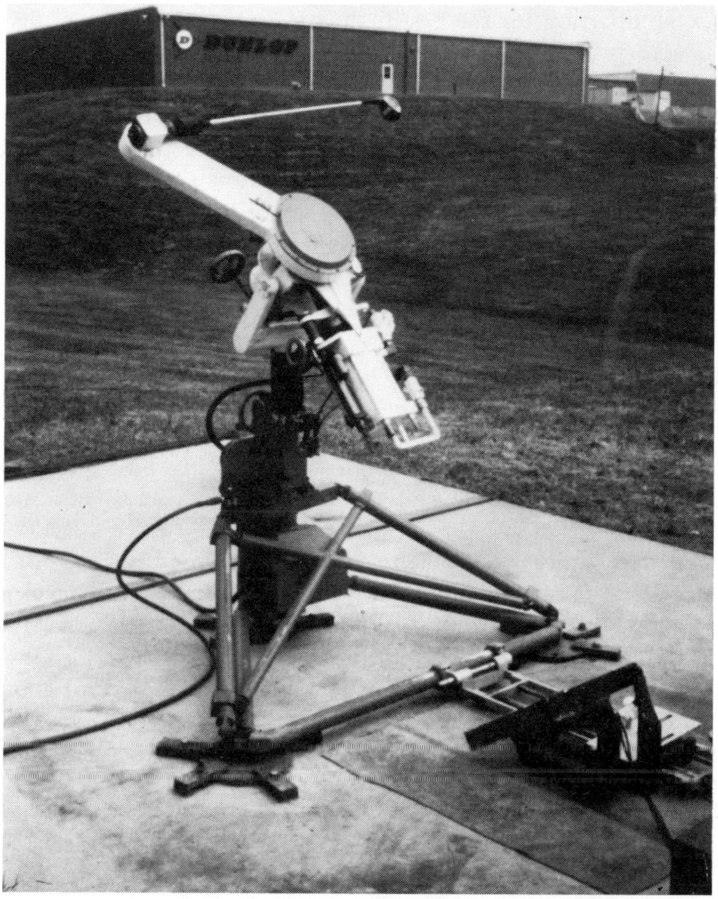


FIG. 12 Machine designed for testing golf balls and clubs. (Courtesy, True Temper Corportation)

The degree to which the boundary layer breaks down to form 'turbulence' behind the ball also affects drag, and all these factors are determined by the type, size and distribution of the indentations in the ball surface.

Nowadays it is normal for golf balls to have circular saucer-shaped indentations or 'dimples' as other shapes tend to have their effect impaired by dirt pick-up, although other shapes have been specified.[22] The diameter, depth, cross-sectional shape and distribution of the dimples is important for the best flight performance in terms of distance and trajectory. Patents have been taken out for crowding dimples as closely together as possible on a ball surface[23] while another particularly specifies that dimples should be omitted from certain areas.[24] It is important, however, that the distribution of the dimples meets the requirements of the ball being 'spherically symmetrical' as referred to in Section 4.2.

## 5 PRODUCT TESTING AND EVALUATION

The performance of tennis, squash and golf balls has to be rigorously tested both in the laboratory and in play. Laboratory testing involves the use of high-speed photography: Fig. 11 shows the impact of a golf club with a ball, indicating the high stresses involved. High-speed photography also allows the assessment of initial trajectory and the spin rate of a golf ball.

Special mechanical equipment is necessary for use in both the laboratory and for field testing, and Fig. 12 shows a mechanical golfer specially developed for testing golf equipment. However, mechanical techniques cannot in any way be a substitute for player testing and rigorous player evaluation of new products must take place. It is only in this way that the more subtle qualities such as 'feel' and 'response' can be properly evaluated, qualities which are essential for the player to exercise his prowess in a successful and satisfying way.

## REFERENCES

1. Stevens, P. H. (1925) BP 230 250.
2. Dunker, H. C. L. (1953) BP 700 544.
3. Dunlop Ltd (1968) BP 1 108 556.
4. Patantex SA and Ciba-Geigy (1978) BP 1 498 732.
5. Colgate Palmolive (1979) BP 2 002 639.
6. Dunlop Ltd (1974) BP 1 367 084.
7. Taylor, D. A. (1976) USP 3 933 967.
8. Industrie Pirelli SPA (1979) BP 2 001 575.
9. General Tire Co. (1978) USP 4 098 504.
10. United States Golf Association (1971) BP 1 225 208.
11. R & A press statement, January 1981.
12. Dunlop Ltd (1964) BP 957 014.
13. Dunlop Ltd (1957) BP 777 959.

14. DuPont Co. (1965) BP 1 011 981.
15. Dunlop Ltd (1967) BP 1 087 566.
16. Questor Corp. (1975) BP 1 386 382.
17. Dunlop Ltd (1975) BP 1 383 422.
18. Bartsch, J. R. (1967) USP 3 313 545; (1969) USP 3 438 933.
19. Spalding, A. G. (1939) USP 2 361 348.
20. Acushnet Co. (1976) BP 1 455 344.
21. Hagopian, G. J. (1970) USP 3 543 380.
22. Uniroyal Inc. (1978) USP 4 090 716.
23. Acushnet Co. (1975) BP 1 415 413.
24. Dunlop Ltd (1978) BP 1 508 039.

# *Chapter* 6

# SHORT FIBRE REINFORCED ELASTOMERIC COMPOSITE MATERIALS

T. ECCERSLEY

*Chemicals and Textiles Division, Department of Industry, London, UK*

## *SUMMARY*

*Most rubbers are reinforced with fillers of a particulate nature. The truly reinforcing fillers such as carbon black or silica reinforce the rubber matrix as a result of the structure or surface activity of the filler particle, though most of these fillers consist of approximately spherical particles. The use of particulate fillers gives equal reinforcement in all three dimensions, hence giving rise to isotropically reinforced rubbers.*

*Little work has been done, however, in the field of fibre reinforcement of rubbers, except in the traditional field of woven fabric reinforcement or braiding, i.e. in belting, hose, tyres, etc. Fibres of suitable length and type are not as easily available as are particulate fillers. The use of discrete length fibres as reinforcement in rubbers should give rise to interesting properties. Theoretical predictions (Guth–Gold equation) indicate that the modulus of a fibre-reinforced rubber should rise rapidly with increasing fibre aspect ratio (length to diameter ratio). There is therefore little point in heavily loading a rubber with fibres as the resultant material will lose the low modulus and high extensibility which characterise rubber. Hence the most promising fibre-reinforced rubbers will arise at the lower fibre loadings. The advantage of using fibre reinforcement as opposed to particulate reinforcement is that by careful control of processing conditions it should be possible to align the fibres so that a composite with marked anisotropy can be produced. This would have distinct advantages in certain applications.*

*Only recently (within the last eight years or so) has much work on fibre-reinforced rubbers been reported. A review in August 1978 described the use of several fibres such as cellulose, carbon, aliphatic and aromatic nylons in such rubbers as natural, nitrile or ethylene–propylene polymers. Previously some work has been reported on the use of carbon fibres in the high-performance field of fluorocarbon rubbers.*

## 1 INTRODUCTION

A composite is defined as a material made up of 'constituents that remain recognisable'. In the polymer field all filler-reinforced or extended materials could thus be regarded as being 'composites'. This title, however, is not generally applied to particulate-modified polymers, only normally being used to refer to fibre-reinforced polymers.

Composite technology is a relatively new field, the two principal journals[1] devoted to its expertise having been in existence only for about 15 years. However, in recent years composites consisting of high-strength reinforcing fibres embedded in a variety of matrix materials have found widespread use and acceptance. In the non-metallics field, much use is made of continuous and discontinuous fibre reinforcement in plastics (e.g. glass-reinforced plastics, GRP) and of continuous fibre (in the form of woven fabric) in rubbers. Discontinuous (short length) fibre has been used only to a very limited extent in rubbers.

The reinforcement of a rubber with fibres would combine the elastic behaviour of the rubber with the strength and stiffness of the reinforcing fibre. Composites of this type would be useful engineering materials. With continuous fibre reinforcement (as in V-belts, tyres and hose) difficulties are experienced in manufacture of the articles. On the other hand, the preparation of an intricate engineering component could be accomplished by using short fibres as the reinforcing medium for a rubber, which could then be processed by the well-known extrusion and transfer or injection-moulding techniques used in the rubber industry.

According to O'Connor,[2] the limitations to date of short fibres in rubber compounding applications have been due to difficulties in achieving uniform dispersion of fibres, fibre breakage during processing, and difficulties in handling, incorporating and bonding the short fibres into the rubber matrix. These problems are more serious with the high viscosities encountered with most elastomers during processing operations, when compared with materials such as fibre-reinforced polyester or epoxy resins. O'Connor further goes on to state that in order to use a fibre-reinforced elastomeric composite most effectively, a basic understanding must be obtained of how the various properties depend on the fibre properties, the matrix properties and the processing methods.

A review by Blackley and Pike[3] in 1974 showed very few literature references to short fibre reinforcement of rubber, whereas four years later (1978) a review by Campbell[4] showed this no longer to be true. Indeed, some elastomeric hoses reinforced with short length cellulose fibres are currently being offered commercially in Europe.[5]

The early use of fibres (e.g. bast fibre) in rubber was merely as a cheapening aid. Later workers have, however, looked at fibres in the hope that they will significantly improve the properties of the rubber. The presence of fibres in a rubber increases the modulus (as does the presence of a reinforcing particulate filler), but other improvements are also sought with fibre addition. The possibility of much improved properties, brought about by alignment of fibres, to give a composite with some measure of anisotropy is of great interest.

The rubber industry has been slow to accept short fibre composites owing to the non-availability of suitable short fibres and also because such materials present two major departures from the habits and experience of rubber workers. Firstly, a composite is generally an engineering material with non-elastic properties (although fabric-reinforced rubber articles such as tyres and hose are certainly flexible), and furthermore, rubber compounds are generally isotropic in nature whereas true composites are mainly anisotropic.

## 2 BACKGROUND TO SHORT FIBRE REINFORCEMENT OF RUBBER

### 2.1 Early Work

As long ago as 1936, Guth, Gimha and Gold[6] derived an equation for calculating the modulus of a fibre-reinforced matrix, applicable to fibre-reinforced rubbers, which has since been much quoted. This equation is commonly referred to today as the 'Guth–Gold equation' and may be expressed as follows:

$$G = G_0\,(1 + 0{\cdot}67fc + 1{\cdot}62f^2c^2)$$

where $G$ = modulus of composite material, $G_0$ = modulus of matrix material, $f$ = length to diameter ratio (aspect ratio) of fibre, and $c$ = volume concentration of fibre. When the fibre aspect ratio is in the range 10–50, moduli ratio of $10^2$–$10^3$ can be achieved if there is good adhesion between fibre and matrix.

Records of any practical applications of fibre-reinforced rubbers are not in evidence for many years after the Guth–Gold paper. A patent in 1951[7] claimed that the improved dispersion of carbon black in rubber could be brought about by the addition of bast fibres from Douglas fir bark. At this period some work was also reported with the natural fibrous filler Solkafloc, where it was used in rubber to reduce shrinkage, kill nerve and act as a processing aid.

### 2.2 Synthetic and Natural Fibres in General-Purpose Rubbers

In more recent years a range of synthetic fibres such as rayon, nylon, polyester, glass, carbon and Kevlar (aromatic polyamide) together with the natural fibres cellulose and asbestos have been dispersed in various rubbers such as natural rubber, styrene–butadiene rubber, polychloroprene–ethylene–propylene terpolymer, polyurethane and fluorocarbon rubbers.

Derringer[8] reported in 1971 that short fibres were then in use as fillers for rubber compounds. He examined nylon, rayon, polyester and acrylic flock cut to approximately 0·4 mm and dispersed in natural rubber. He claimed that 0·4 mm was about the best for optimum dispersion. Work was also done with glass fibres, originally 25 mm in length, but these were found on microscopic examination to have broken (owing to their brittleness) to lengths of 0·1 mm and much less.

Derringer claimed that little improvement of the strength of the rubber occurred unless there was good fibre-to-matrix adhesion. He found good adhesion to be promoted by the use of a hydrated silica–resorcinol–hexamethylene tetramine (HRH) bonding system. Furthermore, Derringer reported that the modulus plots showed rather wide variations amongst the fibre types, especially at high fibre loadings. Rayon was the most reinforcing fibre with respect to modulus at high loadings and acrylic at lower loadings. Nylon apparently imparted the least modulus reinforcement at all loadings, followed by glass fibre. Even the poorer reinforcing fibres, i.e. nylon and glass fibre, however, imparted higher low-strain moduli than high-modulus carbon blacks. Derringer pointed out quite rightly that high modulus at low extension can be very important because many rubber articles do not encounter strains of larger than 100% in service. He also indicated that orientation of the fibre during processing and the subsequent retention of this during moulding of an article could produce anisotropy which could lead to useful improved properties.

**2.3 Natural Cellulose Fibre Composites**

A later paper by Anthoine *et al.*[9] in 1975 dealt mainly with the use of natural short length cellulose fibre as a rubber reinforcement. They claimed that this fibre, being naturally available in short lengths (from wood pulp), should circumvent the complicated route involved in making short fibre from synthetic polymers and thus produce considerable cost savings, the only other naturally occurring fibre, asbestos, currently being unacceptable owing to health hazards.

Previous work by Boustany had shown that for optimum reinforcement the fibre aspect ratio should be in the region of 100–200. The problem was to achieve this in practice, i.e. for filaments such as nylon and polyester, available in diameters of 20–30 $\mu$m, a length of 4–5 mm is needed to give the required aspect ratio. In practice, however, such fibre lengths are too great; they become entangled, making subsequent handling and orientation difficult. On the other hand, as reported by Derringer,[8] glass fibres of diameter 8 $\mu$m are available and these have a high aspect ratio (250) at a length of 2 mm. But, as reported by Derringer, glass fibres break during processing to produce lengths of 0·2–0·4 mm, i.e. to give aspect ratios of 25–50, consequently giving less effective reinforcement.

Anthoine and colleagues then led on in their paper[9] to describe the development of a new short fibre composite, based on the natural short length cellulose fibre previously referred to. They indicated that cellulose had not previously been used in its natural fibre form largely because of difficulties in separating the individual fibres. Anthoine claimed to have overcome this separation problem. Suitably treated fibres (having been treated with a good dispersion and bonding agent) were mixed with a range of standard rubber formulations together with a methylene donor such as hexamethylene tetramine to promote bonding. Orientation of the cellulose fibres during processing was achieved and this gave marked anisotropy to the finished mouldings or extruded products.

The authors reported that the oriented cellulose fibres gave an increase in tensile strength but that the most important changes were in increased modulus and reduced elongation at break (though these changes are typical of those caused by any reinforcing filler in an elastomer). The modulus of the fibre-reinforced materials, as defined by Young's modulus, showed a steady increase with increased fibre concentration. Tensile strength, however, first fell with increasing fibre level and then increased beyond that of the matrix material at higher fibre loadings.

Anthoine and co-workers explained this effect with tensile strength as follows. At low concentration the matrix is not restrained by enough fibres and high strains occur in the matrix at low stresses, causing debonding. The matrix strength is reduced by the debonded fibres. When enough fibres are present, as concentration increases, the matrix is restrained. Eventually, at high concentrations, fibre orientation becomes difficult because the polymer cannot easily flow around the fibres and a reduction in tensile strength occurs again. At present, hose reinforced with these oriented natural short length cellulose fibres is available commercially in Europe.[5]

### 2.4. Important Parameters in Fibre Reinforcement

Many of the above conclusions have been arrived at by other workers in the field. Moghe[10] suggested in 1976 that three parameters should be considered to be most important in effecting fibre reinforcement. These parameters were fibre orientation, fibre aspect ratio and adhesion between fibre and matrix. He indicated that the degree and type of adhesion cannot be estimated quantitatively even though its importance is well recognised. Aspect ratio has a considerable effect on composite properties, hence it is important to conserve fibre length as much as possible during rubber composite processing operations. Moghe suggested, however, that the first parameter listed, that of fibre orientation, had the largest effect on composite properties. During processing (milling, extrusion, etc.) the fibres tend to orient along the flow direction causing mechanical properties to vary in different directions (i.e. marked anisotropy).

In this paper, Moghe went on to describe work carried out with cellulose and Aramid (aromatic polyamide, 'Kevlar') fibres in standard carbon black reinforced acrylonitrile–butadiene and polychloroprene rubbers. All compounds, including dispersion of fibres, were prepared on a two-roll mill. Various combinations of milling time, nip opening and roll-to-roll friction ratio were examined. Moghe found that of these parameters only the mill nip opening influenced the mechanical properties of milled composites. In other words, he suggested that all orientation of fibres occurred during the first pass through the mill nip and further milling did not improve matters.

**2.5 Fibre Orientation and Bonding Agents in Composites**

The determination of fibre orientation in an elastomer composite gave rise to other research programmes. Foldi[11] reported on work he had carried out with a range of fibres in a range of standard rubbers. He attempted to measure the degree of orientation of fibres in the composite by tensile tests in the 'machine direction', i.e. parallel to the direction of probable orientation of fibres due to flow during processing, and in the 'cross-machine' direction, i.e. at right angles to the direction above. These tensile properties were measured on 'green' (uncured) sheets of composite and gave a fairly good indication of the amount of fibre orientation (Foldi indicated that 100% orientation was never achieved). Examination of composite sections by light microscope gave only a poor indication of fibre orientation; it was very difficult to pick out the fibres owing to the carbon black reinforced matrix. Foldi reported that, of the fibres examined, nylon and aromatic polyamide (Aramid K) gave the best improvement in green strength, producing a 27–29-fold increase over the control stock at only 10 pphr fibre loading. Vulcanised tensile properties were also much modified by the polyamide fibres and Foldi attributed this to the unbroken, flexible fibres, still apparently having aspect ratios of the order of 200–500 in the composite, compared with other fibres such as glass which broke into short fragments during incorporation into the matrix.

The fibres previously referred to, such as glass, cellulose, rayon and Aramid, were examined in work reported by O'Connor.[2] He also examined the more exotic material, carbon fibre. These fibres were incorporated into typical general-purpose rubber compounds. Typical comments on orientation of fibres by processing operations were made. O'Connor, however, also investigated the use of bonding agents to attempt to promote fibre–matrix adhesion. As did Derringer,[8] O'Connor used a hydrated silica–resorcinol–hexamethylene tetramine (HRH) system to promote bonding. He did indeed modify the fibre-to-matrix bond as indicated by the quoted tensile properties. Young's modulus in particular was increased by the HRH system for most fibres, though glass and carbon fibres showed little improvement.

O'Connor attributed the poor performance of these latter fibres to the reduction in fibre length during processing. He suggested, as others have done, that an aspect ratio of between 40 and 250 is considered necessary for successful fibre reinforcement of an elastomer.

**2.6 Carbon Fibre Composites**

Little work has been reported on the use of carbon fibres as the reinforcing medium for elastomeric composites. Gill[12] in his book referred in 1972 to the use of carbon fibre in high-performance rubbers such as fluorocarbons. He quoted three authors, as follows.

Genin[13] reported success with carbon fibres in rubbers to give improved temperature performance. Particular applications quoted were in fluorocarbon copolymers. Bergstrom[14] also reported improvements in these materials, particularly in tensile strength measured at 120 °C. However, only limited detail is available of these test results in both Bergstrom's and Genin's papers as the majority of their work was on other aspects of fluorocarbon polymers, with carbon fibre reinforcement as a side-issue.

Work specifically on the use of carbon fibres in fluorocarbon rubber[15,16] and ethylene–propylene (EPDM) and isoprene–isobutylene (butyl) rubbers[16] has been reported by Sieron. Carbon fibre was shown to improve the tensile strength retention after ageing at 250 °C to 30% compared with 15% retention over the same ageing period without carbon fibre reinforcement. No indication of final fibre length after processing is indicated in the fluorocarbon paper, neither is there any mention of the importance of fibre orientation and fibre–matrix bonding. It must be remembered, however, that this work was reported in 1963 and is thus earlier than a lot of the work described earlier in this chapter. Sieron also describes, in his second paper,[16] work on carbon fibre reinforcement of butyl rubber and ethylene–propylene (EPDM) rubber. Carbon fibre was again used in an attempt to improve the high-temperature strength of the rubbers. Other rather exotic fibres were examined also, such as fibrous potassium titanate and magnesium silicate, as was asbestos and polyimide fibre. The most significant improvements were in hot tear strength where values 60% greater than those without carbon fibres were achieved in fluorocarbon rubbers. Carbon fibre also improved the heat-ageing properties of the lower performance butyl and EPDM rubbers. Again, no mention is made of the importance of fibre aspect ratio, fibre orientation or fibre–matrix adhesion.

### 2.7 Current Position

Campbell (1978) published a review paper on 'Short Fibre Reinforcement of Rubber'.[4] This paper covers most of the other references in this chapter and provides a very up-to-date picture of the field. The major current usage of short length fibres in rubber reinforcement appears to be that of cellulose offered by Monsanto Ltd. A recent paper by Goettler *et al.*[17] describes the manufacture of hose using cellulose fibre as the reinforcing medium. Careful control of processing conditions and extruder die design allows the alignment of the short cellulose fibres in preferred directions. The paper shows how axially oriented fibres can be used to reinforce circumferentially hose for low to intermediate pressure applications. A further recent paper, by Shen *et al.*,[18] discusses the large-scale mixing of cellulose fibre-reinforced rubbers.

## 3 THEORETICAL TREATMENT OF REINFORCEMENT OF A MATRIX

### 3.1 Manson and Sperling's Treatment

Much theoretical treatment of the reinforcement of a matrix by a spherical filler particle has been reported and will not be discussed here. However, as the shape of a filler particle is elongated from a spherical shape to a fibre, the filler stiffens the composite to a greater extent and begins to carry increasing fractions of the load.[19]

It can be shown that a matrix can transfer a major portion of an applied stress to fibrous elements only if the ratio of length $l$ to diameter $d$ exceeds a critical value $l_c$,[20] where

$$\frac{l_c}{d}=\frac{\sigma_f}{2\tau}$$

where $\sigma_f$ is the maximum stress in the fibre and $\tau$ is the matrix or interface shear strength. The fibre ends do not carry load; as the fibre length increases, the relatively ineffective portions of the fibre decrease. The average stress in a fibre $\bar{\sigma}_f$ is given by

$$\bar{\sigma}_f=\sigma_f\left(1-\frac{l_c}{2l}\right)$$

Substitution of typical values in the above equations shows that fibre would need to have a ratio of $l_c/d$ (i.e. aspect ratio) of about 60 in order to load the fibre to its maximum stress at the centre. Hence, any fibre aspect ratio greater than this should behave as though it were a continuous fibre as far as load bearing is concerned.

### 3.2 Nielsen and Chen's Treatment

With random fibres in a plane a modified rule of mixtures has been proposed by Nielsen and Chen[21] to predict the modulus of the composite:

$$G = KG_f c + G_0 (1 - c)$$

where $K$ is an empirical fibre efficiency parameter which depends on the volume fraction of fibre $c$ and on the ratio of fibre to matrix modulus $G_f/G_0$. For $100 > G_f/G_0 > 10$ and $0{\cdot}6 > c > 0{\cdot}1$, $K$ lies between 0·15 and 0·56.

Other authors, for example Tsai and Halpin, have suggested more complex versions of the Nielsen treatment. The reader is referred to Chapter 12 of Manson and Sperling's excellent book[19] on the subject.

### 3.3 Guth–Gold Equation

The Guth-Gold equation is usually described as:

$$G = G_0 (1 + 0{\cdot}67fc + 1{\cdot}62f^2 c^2)$$

where $G$ = modulus of composite material, $G_0$ = modulus of matrix material, $f$ = aspect ratio of fibre, and $c$ = volume concentration of fibre. This equation has already been dealt with briefly (Section 2.1) where it was shown how it could be used to predict the modulus of a fibre-reinforced composite. Later it will be seen how the Guth–Gold equation can be applied to the modulus of a composite.

## 4 MATERIALS USED IN THE PREPARATION OF EXPERIMENTAL COMPOSITES

Much work has been reported over a period of many years on the use of particulate reinforcing agents in elastomeric materials. Little work, however, has been done with short length fibre reinforcing agents. This is in contrast with the plastics field where much use is made of short and long length fibres in products such as glass-reinforced plastics.

A survey of the use of fibres as reinforcement in rubber was carried out. It was found that some work had been done on the use of natural and the cheaper synthetic fibres in several general-purpose rubbers, and work in this area appeared at present to be accelerating and even reaching commercialisation. The author, however, felt—and this was confirmed by professional colleagues—that some useful study could be made of the reinforcement of the rather exotic material, silicone rubber, with a rather exotic fibre, namely carbon fibre, in the hope that the high performance of this rubber and this fibre could be married to produce a novel composite material. Consequently it was proposed that a range of carbon fibre–silicone composites should be prepared and an examination made of such effects as alignment of the fibres, fibre length, fibre content, chemical treatment of the fibre and comparison with one other fibre (glass). Conclusions would be drawn from the experimental findings and suggestions would be made for future work.

### 4.1 Silicone Polymer

Silicone polymers are prepared from substituted silanes. For example, dimethydichlorosilane can be reacted with water to form dimethylsilanol:

$$\mathrm{Cl{-}\underset{CH_3}{\overset{CH_3}{\underset{|}{\overset{|}{Si}}}}{-}Cl + 2H_2O \rightarrow 2HCl + HO{-}\underset{CH_3}{\overset{CH_3}{\underset{|}{\overset{|}{Si}}}}{-}OH}$$

which is not stable but reacts with other silanol molecules to eliminate water:

$$\mathrm{HO{-}\underset{CH_3}{\overset{CH_3}{\underset{|}{\overset{|}{Si}}}}{-}OH + HO{-}\underset{CH_3}{\overset{CH_3}{\underset{|}{\overset{|}{Si}}}}{-}OH \rightarrow HO{-}\underset{CH_3}{\overset{CH_3}{\underset{|}{\overset{|}{Si}}}}{-}O{-}\underset{CH_3}{\overset{CH_3}{\underset{|}{\overset{|}{Si}}}}{-}OH + H_2O}$$

The disilanol continues to react under production conditions to form cyclic polydimethylsiloxanes.

The silicone polymer employed for this project was 'Gum 200' supplied by Dow Corning Ltd. It is essentially of the above chemical composition with vinyl end-groups. It is reported to have a weight-average molecular weight of between $8 \times 10^5$ and $1 \times 10^6$ and a dispersity ($M_w/M_n$) of between 5 and 10. This polymer was chosen because it was easily classifiable.

### 4.2 Silica Reinforcing Filler

A silica filler was used to reinforce the silicone polymer to provide a fairly strong matrix for the composites. 'Aerosil R972', supplied by Degussa, was selected for this purpose. 'Aerosil' is a coagulated silicon dioxide prepared by the hydrolysis of silicon tetrachloride in an oxygen–hydrogen flame. It consists of very fine spherical particles of 10–20 nm ($10–20 \times 10^{-9}$ m) average diameter. Most 'Aerosil' silicas are hydrophilic in nature and interact strongly with certain polymers such as silicones. However, by reaction with a chemical such as dimethyldichlorosilane, the active silanol groups of the silica can be made hydrophobic. 'Aerosil R972', the silica used in this project, had 70% reactive silanol groups blocked by the silane to make it an organophilic silica. It possesses a surface area (BET method) of about 120 $m^2/g$ and produces reasonable reinforcement of a silicone polymer when mixed into the polymer at the level of 40 parts by weight per 100 of polymer and crosslinked with a suitable peroxide.

### 4.3 Carbon Fibre

Carbon fibre is made from a polymeric precursor such as polyacrylonitrile by progressive thermal degradation whilst held under tension, first at about 220 °C and later up to a temperature of 1600 °C. Material of this type was developed at the Royal Aircraft Establishment, Farnborough, UK, and consists of a few different types depending on the detailed thermal history of the material. All have Young's moduli of the order of 200–500 GPa.

The particular carbon fibre selected for this project was produced commercially by Courtaulds Ltd, described as 'Grafil HTS', which indicates that it is the high tensile strength version and has been surface treated (acid etched) to improve bonding to a matrix. The carbon fibre had a diameter of 8 $\mu$m (as most do) and was obtained in chopped lengths of approximately 3 mm (in fibre bundles of 10 000).

### 4.4 Peroxide Crosslinking Agent

This was a standard one for silicone rubbers, namely 2,4-dichlorobenzoyl peroxide, sold under several trade names, one of which is 'Perkadox PDS-40'. It is employed at levels of 2% (on polymer) upwards and crosslinks the polymer via the methyl groups.

### 4.5 Silane Coupling Agents

These were applied in water suspension to samples of carbon fibre, according to procedures laid down in Union Carbide and Dow Corning Trade literature,[22,23] at levels of 1 part (by weight) active silane to 5 parts of carbon fibre.

# 5 INTERPRETATION OF EXPERIMENTAL RESULTS

## 5.1 Young's Modulus and the Guth–Gold Equation

Young's Modulus was determined on a range of composites containing Grafil HTS carbon fibre, originally cut to 3 mm in length. The composites examined consisted of (parts by weight):

| | |
|---|---|
| Silicone Gum 200 | 100 |
| Silica R972 | 40 |
| 2, 4-Dichlorobenzoyl peroxide | 4 |

together with carbon fibre at the levels of zero (control sample), 5, 10, 15, 20, 25 and 30 parts by weight per hundred of silicone polymer (these representing respectively 0, 0·02, 0·04, 0·058, 0·076, 0·093 and 0·110 volume fraction of carbon fibre).

The values of Young's modulus for each composite were divided by the Young's modulus of the control (zero carbon fibre content, i.e. the matrix material) to give modulus ratios. These values are compared (Table 1) with values calculated according to the Guth–Gold equation previously referred to. Values of aspect ratio fed into the Guth–Gold equation have been selected so that the calculated values of modulus ratio straddled the experimentally determined values. It will be seen by reference to Table 1 that at the lowest volume fraction (0·02) of carbon fibre the Guth–Gold prediction suggests a fibre aspect ratio of about 37, i.e. a fibre of 0·30 mm in length. At the upper end of the fibre loading range, at a volume fraction of 0·110 carbon fibre, the experimental value fits an aspect ratio of about 14, i.e. a fibre of 0·11 mm in length.

Microscopic examination of the carbon fibres carefully removed (by solvent extraction of the polymer) from a non-vulcanised composite containing 0·058 volume fraction (15 pphr) of fibre was then carried out. A typical microphotograph (Fig. 1) shows the composite to contain a range of fibre lengths. This may be compared with a microphotograph of the original carbon fibre (Fig. 2) which shows no short length debris. Size distribution of the extracted fibres calculated from individual measurements on a typical microphotograph is shown in Fig. 3. This indicates a weighted average fibre length of 0·19 mm, i.e. an aspect ratio of 24. This is of the same order as the values predicted by the Guth–Gold equation, i.e. aspect ratio between 14 and 37. It must be remembered, however, that, as mentioned earlier, the Guth–Gold equation does not take into account the modulus of the fibre itself nor does it assume anything other than perfect matrix-to-fibre adhesion.

TABLE 1

RATIOS OF MODULI ($G/G_0$) CALCULATED ACCORDING TO THE GUTH–GOLD EQUATION[a] COMPARED WITH RATIOS DERIVED FROM YOUNG'S MODULUS MEASUREMENTS ON CARBON FIBRE–SILICONE COMPOSITES

| *Calculated values* | | *Volume fraction (pphr)* | | | | | |
|---|---|---|---|---|---|---|---|
| *Aspect ratio* | *Length (mm)* | *0·02 (5)* | *0·04 (10)* | *0·058 (15)* | *0·076 (20)* | *0·093 (25)* | *0·110 (30)* |
| 12·5 | 0·10 | 1·27 | 1·74 | 2·34 | 3·10 | 3·97 | 4·98 |
| 13·75 | 0·11 | 1·31 | 1·86 | 2·56 | 3·47 | 4·51 | 5·72 |
| 15·0 | 0·12 | 1·35 | 1·98 | 2·81 | 3·87 | 5·09 | 6·52 |
| 16·25 | 0·13 | 1·39 | 2·12 | 3·07 | 4·30 | 5·71 | 7·37 |
| 17·5 | 0·14 | 1·43 | 2·26 | 3·35 | 4·76 | — | — |
| 18·75 | 0·15 | — | 2·83 | — | — | — | — |
| 37·5 | 0·30 | 2·40 | — | — | — | — | — |
| Experimentally determined values of $G/G_0$ | | 2·38 | 2·61 | 3·18 | 4·90 | 5·03 | 5·94 |

[a]See text.

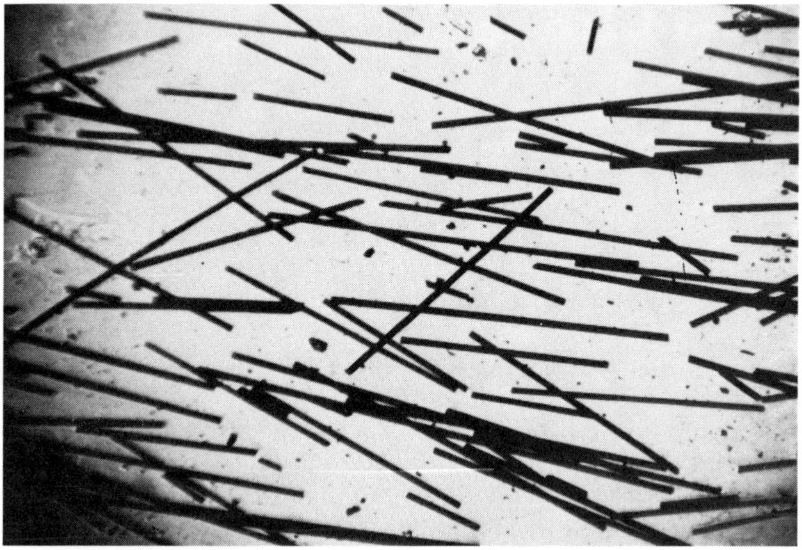

FIG. 1. Carbon fibres extracted from a composite (120×).

FIG. 2. Microphotograph of original carbon fibres before incorporation into composites (120×).

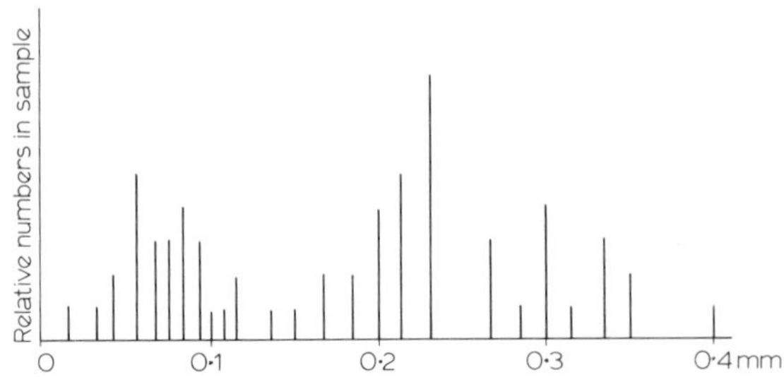


FIG. 3. Size distribution of extracted carbon fibre.

### 5.2 Vulcanisate Tensile Properties with Fibre Aligned in Direction of Testing

The composites referred to in the previous section were also subjected to a standard rubber tensile properties determination. Prior to these tests, Monsanto oscillating disc rheometer tests were carried out on unvulcanised samples. The results of all these tests are shown in Table 2. The rheometer test indicated that the viscosity of the composite (as expressed by minimum torque value) increased regularly with increase in fibre content. Increasing fibre content did not, however, affect the level of vulcanisation (as expressed by the torque change) or the rate of vulcanisation (as expressed by the 90% cure time values).

Addition of carbon fibre initially caused a progressive decrease in the elongation at break of the composites. At 0·02 volume fraction (VF) the modulus (tensile stress) values were lower than those of the matrix material. At higher loadings the tensile moduli increased with fibre content up to 0·076 VF. Above this apparently critical level the modulus decreased again and the elongation at break increased. Plots of the stress–strain curves up to 300% elongation for the range of composites (Fig. 4) show that all composites deviated from the almost straight-line response of the non-fibre-reinforced matrix. The 0·076 VF (20 pphr) composite showed the optimum reinforcement. Beyond this level there appeared, after an initial high stress level, a long flat plateau on the stress–strain curve. Visual examination of the 0·093 and 0·110 VF composites during stress–strain measurements on the Instron Tensometer showed that considerable flow and necking of the sample was occurring under the stress. This suggests that, at the higher level of fibre concentration, there was pronounced loss of fibre-to-matrix adhesion. The upper end of the stress–strain curve for these two composites approximates to that for the control matrix material, hence it

TABLE 2

EFFECT OF CARBON FIBRE LEVEL ON MIXED RUBBER RHEOLOGICAL AND VULCANISATE PROPERTIES

| Base mix: | Silicone Gum 200 | 100 parts by weight |
|---|---|---|
| | Silica R972 | 40 parts by weight, pphr |
| | 2,4-Dichlorobenzoyl peroxide | 4 parts by weight, pphr |

| | | *Carbon fibre 'Grafil HTS', chopped, volume fraction (pphr)* | | | | | | |
|---|---|---|---|---|---|---|---|---|
| | | *0* | *0·02 (5)* | *0·04 (10)* | *0·058 (15)* | *0·076 (20)* | *0·093 (25)* | *0·110 (30)* |
| Rheological properties, Monsanto oscillating disc rheometer | Torque (in lb, min.) | 9·0 | 10·4 | 12·0 | 14·4 | 15·8 | 17·6 | 22·5 |
| | Torque (in lb, max.) | 16·0 | 17·3 | 19·0 | 21·2 | 23·3 | 24·6 | 30·4 |
| | Torque change | 7·0 | 6·9 | 7·0 | 6·8 | 7·5 | 7·0 | 7·9 |
| | 90% cure time | 2·0 | 1·8 | 2·0 | 1·9 | 2·0 | 1·9 | 1·9 |
| Vulcanisate tensile properties, parallel to fibre alignment, (cure 10 min at 120 °C) | Tensile strength (MPa) | 8·05 | 6·44 | 6·45 | 4·92 | 5·45 | 4·20 | 4·27 |
| | Elongation at break (%) | 710 | 630 | 590 | 520 | 470 | 510 | 460 |
| | 100% modulus (MPa) | 1·03 | 0·96 | 1·17 | 1·36 | 2·10 | 1·67 | 1·83 |
| | 300% modulus (MPa) | 2·93 | 2·60 | 2·67 | 2·79 | 3·73 | 2·70 | 2·86 |
| | 500% modulus (MPa) | 5·20 | 4·89 | 4·80 | 4·65 | — | 4·09 | — |
| | Young's modulus at 10% elongation (MPa) | 2·06 | 4·91 | 5·38 | 6·56 | 10·10 | 10·36 | 12·24 |
| | Tear strength (kN/mm) | 17·7 | 8·4 | 7·3 | 7·9 | 9·5 | 6·0 | 7·0 |

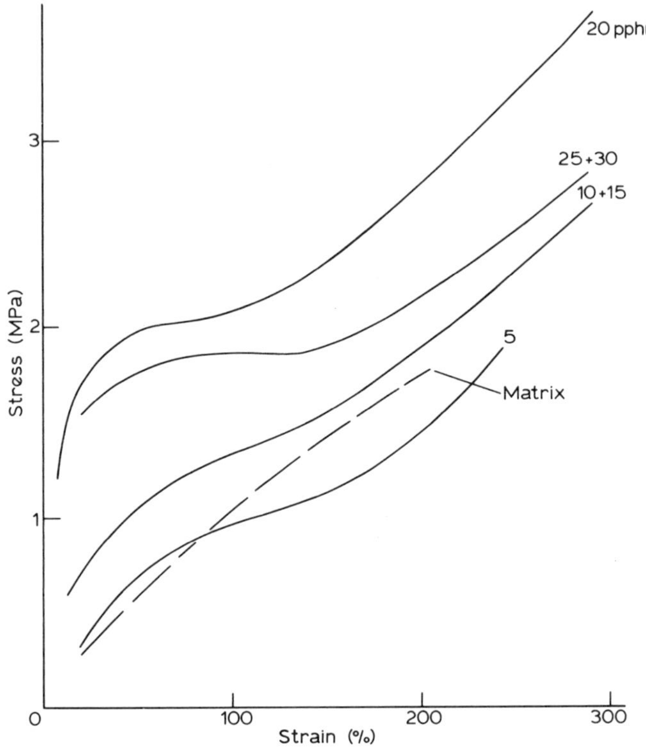


FIG. 4. Stress–strain curves.

would appear that no fibre reinforcement occurs after internal break-down of the fibre–matrix bond.

Tear strength tests on the range of components showed (Table 2) a rather random pattern with increased fibre loading; all composites, however, had tear strengths lower than that of the matrix.

**5.3 Comparison of Aligned and Randomly Oriented Fibres**

Composites were made by plying up thin sheets (0·2–0·5 mm) of unvulcanised material with the fibre orientation in alternate layers at right angles to adjacent layers, thus producing what might be considered a 'randomly oriented' composite. Tensile properties were determined on these random composites at fibre loadings of 0·04, 0·058 and 0·076 VF (5, 10 and 15 pphr) and compared (Table 3) with the previously reported data on aligned composites. Modulus values at 300% elongation were

TABLE 3

EFFECT OF FIBRE ALIGNMENT ON VULCANISATE TENSILE PROPERTIES AT INTERMEDIATE VULCANISATION LEVEL

| Base mix: | | |
|---|---|---|
| | Silicone Gum 200 | 100 parts by weight |
| | Silica R972 | 40 parts by weight, pphr |
| | 2,4-Dichlorobenzoyl peroxide | 4 parts by weight, pphr |

| | | *Chopped carbon fibre 'Grafil HTS', volume fraction (pphr)* | | | | | | |
|---|---|---|---|---|---|---|---|---|
| | | *0* | *0.04 (10)* | | *0.058 (15)* | | *0.076 (20)* | |
| *Fibre orientation with respect to direction of testing* | | — | *Random* | *Parallel* | *Random* | *Parallel* | *Random* | *Parallel* |
| Rheological properties, Monsanto oscillating disc rheometer | Torque (in lb, min.) | 9·0 | 12·0 | | 14·4 | | 15·8 | |
| | Torque (in lb, max.) | 16·0 | 19·0 | | 21·2 | | 23·3 | |
| | Torque change | 7·0 | 7·0 | | 6·8 | | 7·5 | |
| | 90% cure time | 2·0 | 2·0 | | 1·9 | | 2·0 | |
| Vulcanisate tensile properties, parallel to fibre alignment (cure 10 min at 120 °C) | Tensile strength (MPa) | 9·31 | 4·42 | 6·45 | 4·37 | 4·92 | 5·18 | 5·45 |
| | Elongation at break (%) | 710 | 610 | 590 | 540 | 520 | 470 | 470 |
| | 100% modulus (MPa) | 1·03 | 1·07 | 1·17 | 1·56 | 1·36 | 2·39 | 2·10 |
| | 300% modulus (MPa) | 2·63 | 2·03 | 2·67 | 2·48 | 2·79 | 3·42 | 3·73 |
| | 500% modulus (MPa) | 5·30 | 3·54 | 4·80 | 4·07 | 4·65 | — | — |

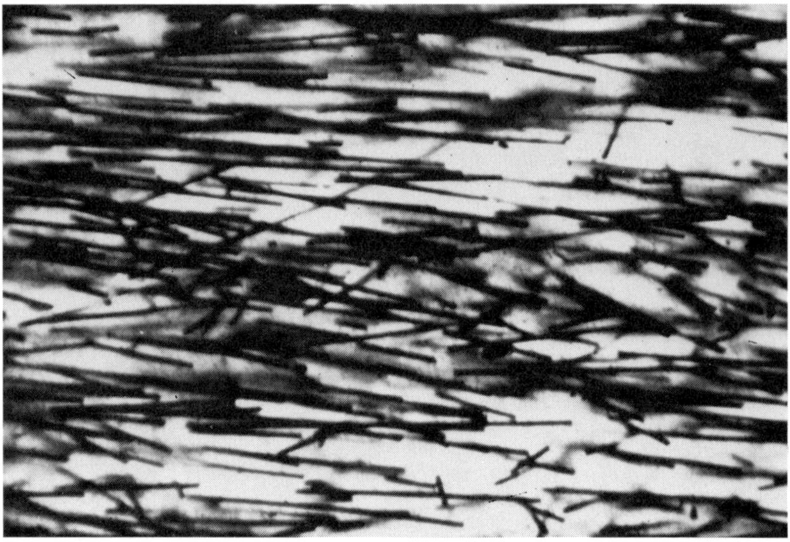

FIG. 5. Composite containing 0·04 VF (10 pphr) aligned carbon fibre: longitudinal section through tensile bar, bar cut from sheet in machine direction (120 ×).

higher, and elongation at break lower, at all loadings for the aligned composite compared with the 'random' samples; this suggests that the orientation of the fibre produced better reinforcement of the matrix. Microphotographs were prepared of these aligned and random composites from sections cut in two directions at right angles to each other. At a fibre loading of 0·04 VF (10 pphr) the fibre can be seen (Fig. 5) to be aligned in the direction of milling, though with some fibres lying at an angle to this direction. A cross-section of this material (Fig. 6) shows considerable fibre ends. A sample of the 0·076 (20 pphr) aligned composite was photographed before stress–strain measurement (Fig. 7); a sample was also taken from a test piece which had been strained by 300% (Fig. 8). The unstressed composite showed, as did the 0·04 VF sample, some fibres lying out of line whereas with the stressed sample there was little evidence of any non-aligned fibres.

A longitudinal section of a random composite was microphotographed (Fig. 9). There it can be seen that some fibres are aligned in the direction of the section (bottom of plate) and at right angles to the direction of section (top of plate). A cross-section of this random composite (Fig. 10) confirms this finding.

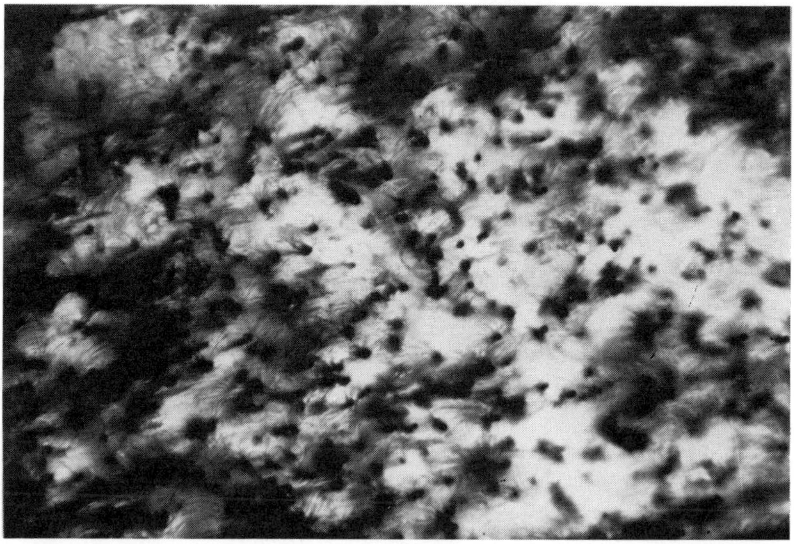

FIG. 6. Composite containing 0·04 VF (10 pphr) aligned carbon fibre: cross-section through tensile bar, bar cut from sheet in machine direction (120 ×).

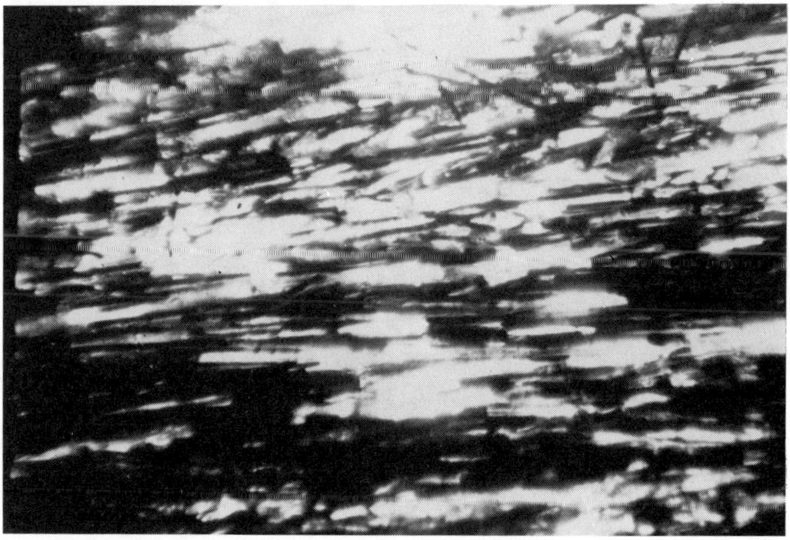

FIG. 7. Composite containing 0·076 VF (20 pphr) aligned carbon fibre: longitudinal section through tensile bar, bar cut from sheet in machine direction (120 ×).

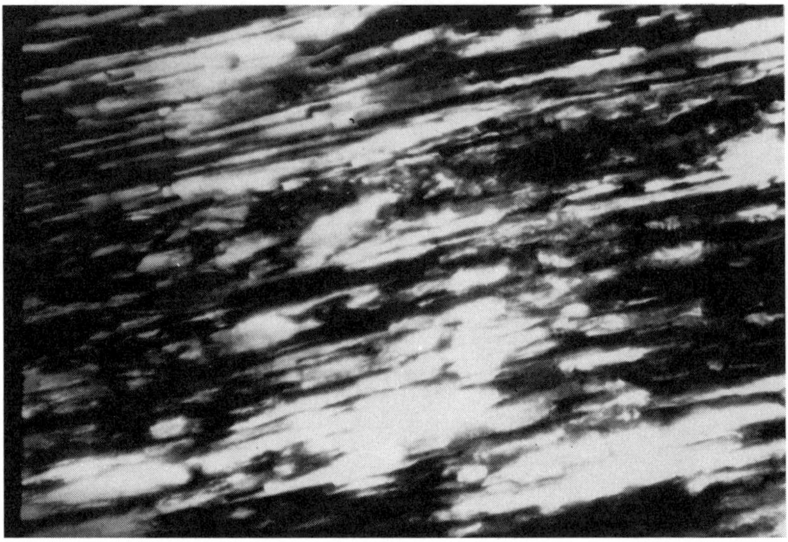

FIG. 8. Composite as in Fig. 7 but after stretching by 300% (120 ×).

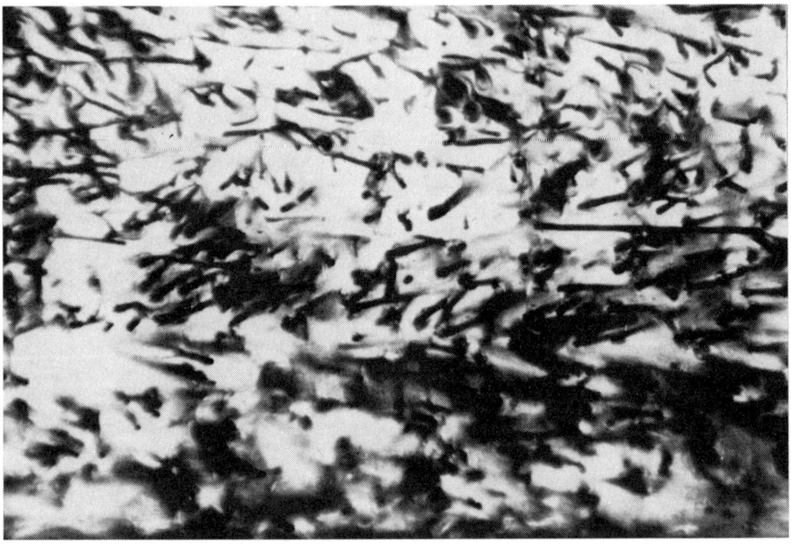

FIG. 9. Composite containing 0·04 VF (10 pphr) random carbon fibre: longitudinal section through tensile bar (120 ×).

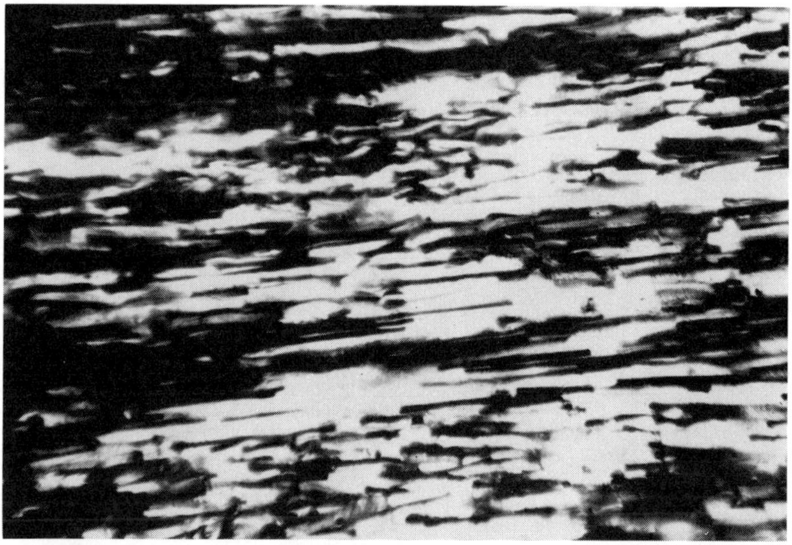

FIG. 10. Composite containing 0·04 VF (10 pphr) random carbon fibre: cross-section through tensile bar (120×).

### 5.4 Comparison of Composites with Fibres Aligned Parallel to and Perpendicular to Direction of Test

Further composites were prepared, this time from the following mix recipe (parts by weight):

| | |
|---|---|
| Silicone Gum 200 | 100 |
| Silica R972 | 40 |
| 2,4-Dichlorobenzoyl peroxide | 8 |
| 'Grafil HTS' carbon fibre, 3 mm | 15 |

The peroxide level was increased to 8 pphr to see how this influenced the state of cure and resultant physical properties. Test sheets were moulded from this 0·058 VF (15 pphr) carbon fibre composite and tensile dumbbells cut so that some had the fibres aligned along the length of the test piece and some had the fibres aligned across the width of the test piece, thus allowing an examination of anisotropy in the composite. Test results showed (Table 4) a considerably higher tensile modulus, but a fairly similar elongation at break was developed with the in-line fibres compared with the cross-direction fibre composites. Microphotographs of these two samples show the orientation of the fibres (Figs. 11–14).

TABLE 4

ANISOTROPY OF 0·058 VF (15 pphr) COMPOSITE

| | *Carbon fibre orientation with respect to direction of testing:* | *Parallel* | *Perpendicular* |
|---|---|---|---|
| Rheological properties, Monsanto oscillating disc rheometer | Torque (in lb, min.) | 14·7 | |
| | Torque (in lb, max.) | 29·8 | |
| | Torque change | 15·1 | |
| | 90% cure time | 1·8 | |
| Vulcanisate tensile properties (cure 10 min at 120°C) | Tensile strength (MPa) | 7·67 | 6·34 |
| | Elongation at break (%) | 560 | 540 |
| | 100% modulus (MPa) | 1·52 | 1·13 |
| | 300% modulus (MPa) | 3·36 | 2·80 |
| | 500% modulus (MPa) | 6·57 | 5·46 |

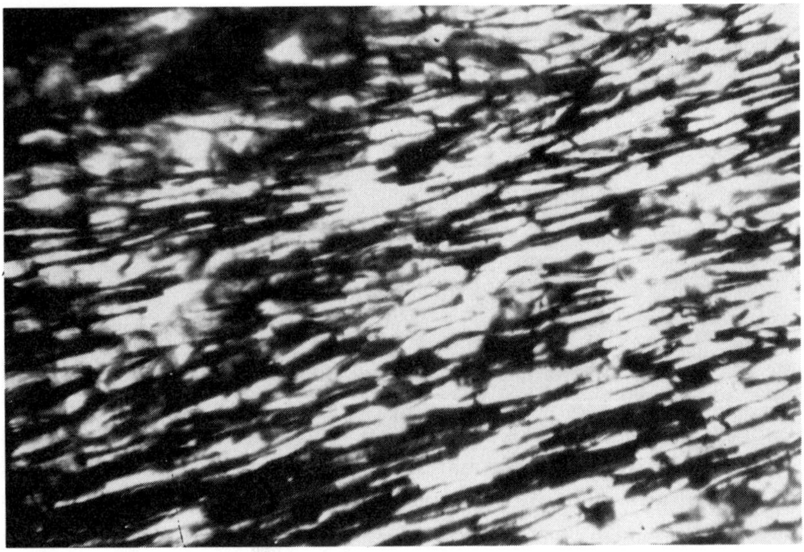

FIG. 11. Composite containing 0·58 VF (15 pphr) aligned carbon fibre: longitudinal section through tensile bar, bar cut from sheet in machine direction (120 ×).

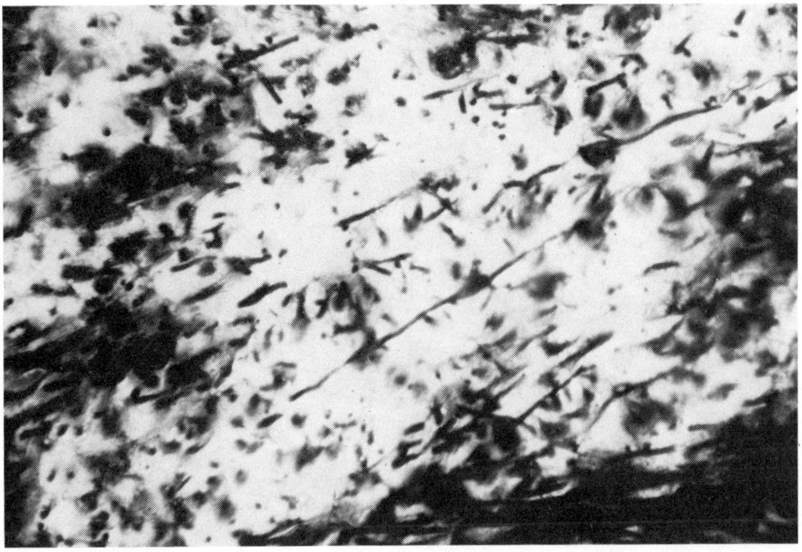

FIG. 12. Composite containing 0·058 VF (15 pphr) aligned carbon fibre: cross-section through tensile bar, bar cut from sheet in machine direction (120 ×).

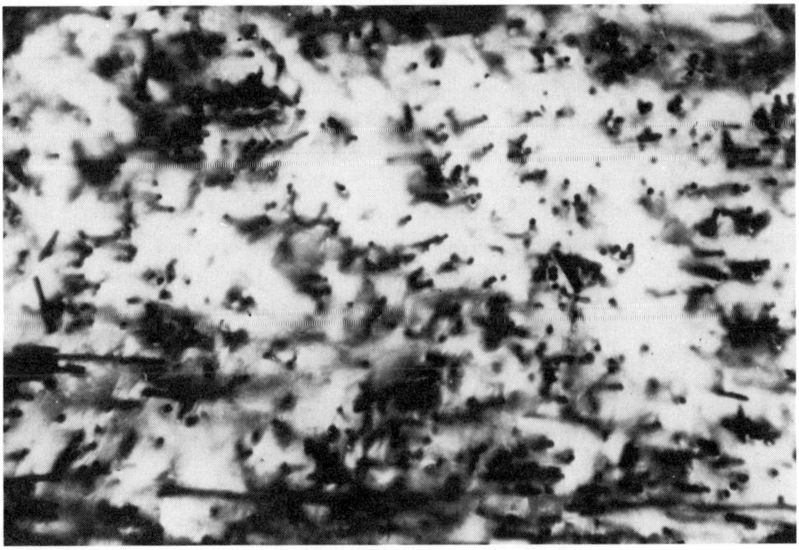

FIG. 13. Composite containing 0·058 VF (15 pphr) aligned carbon fibre: longitudinal section through tensile bar, bar cut from sheet in cross-machine direction (120 ×).

FIG. 14. Composite containing 0·058 VF (15 pphr) aligned carbon fibre: cross-section through tensile bar, bar cut from sheet in cross-machine direction (120 ×).

### 5.5 Effect of Increased Milling Time on Carbon Fibre Composites

All previous composites had been prepared by a standard procedure whereby, after full incorporation of the carbon fibre into the matrix, a further 5 min milling ('alignment time') was given. Moghe[10] indicated that orientation of the fibres could be increased by longer milling times, so samples of unvulcanised composites were milled for 15 and 30 min after completion of fibre incorporation. Tensile properties determined on vulcanisates prepared from these materials (Table 5) showed that reinforcement of the matrix was improved by extra milling. Microphotographs of the composite (Figs. 15 and 16) prepared after 15 min extra milling showed well-oriented fibres.

### 5.6 Chemical Treatment of Carbon Fibre Surface in an Attempt to Increase Fibre-to-Matrix Bond Strength

Samples of the short length carbon fibre were treated with a range of silane coupling agents in the hope that these would improve the fibre-to-matrix bond (evidence of weak bond). Since carbon fibre is known to have an inert surface, however (though little is apparently known of its surface chemistry), it was not really expected that much improvement

TABLE 5
EFFECT OF MILLING TIME ON PROPERTIES OF CARBON FIBRE–SILICONE COMPOSITE

| Base mix: | | |
|---|---|---|
| | Silicone Gum 200 | 100 parts by weight |
| | Silica R972 | 40 parts by weight |
| | Carbon fibre 'Grafil HTS' | 15 parts by weight |
| | 2,4-Dichlorobenzoyl peroxide | 4 parts by weight |

| | | *Alignment time*[a] *on mill (min)* | | |
|---|---|---|---|---|
| | | *5* | *15* | *30* |
| Vulcanisate tensile properties (cure 10 min at 120 °C) | Tensile strength (MPa) | 6·44 | 5·76 | 6·22 |
| | Elongation at break (%) | 630 | 570 | 550 |
| | 100% modulus (MPa) | 0·96 | 1·57 | 1·80 |
| | 300% modulus (MPa) | 2·60 | 2·87 | 3·33 |
| | 500% modulus (MPa) | 4·87 | 4·95 | 5·55 |

[a]Alignment time = extra milling time after complete incorporation/dispersion of fibres.

could be made by silane treatment. Tests results, however (Table 6), showed that of the silanes examined, three types increased the Young's modulus of the composite by a factor of over two, though with little effect on the tensile modulus (stress at given elongations). Thus some improvement could be brought about by the use of a silane coupling agent.

## 5.7 Comparison of Silica-Reinforced and Silica–Carbon Fibre Reinforced Materials

Vulcanisates were prepared from mixes containing increased amounts of 'Aerosil' silica R972 and compared with some of the previously described carbon fibre composites. Test results showed (Table 7) that silica alone as reinforcing filler gave poorer processability (as indicated by increased value of minimum rheometer torque) with lower values of Young's modulus, hence carbon fibre has an advantage over extra silica as a reinforcing filler.

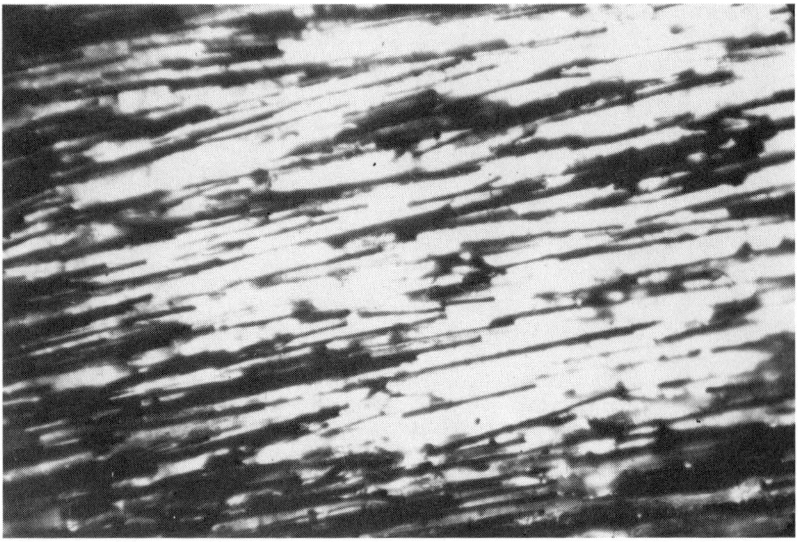

FIG. 15. Composite containing 0·058 VF (15 pphr) aligned carbon fibre: longitudinal section through tensile bar, bar cut from sheet in machine direction, increased (15 min) alignment time (120 ×).

FIG. 16. Composite containing 0·058 VF (15 pphr) aligned carbon fibre: cross-section through tensile bar, bar cut from sheet in machine direction, increased (15 min) alignment time (120 ×).

TABLE 6

SILANE TREATMENT OF GRAFIL HTS CARBON FIBRE

| Base mix: | | |
|---|---|---|
| | Silicone Gum 200 | 100 parts by weight |
| | Silica R972 | 40 parts by weight |
| | Grafil HTS | 10 parts by weight |
| | 2,4-Dichlorobenzoyl peroxide | 8 parts by weight |

| | | Silane[a] | | | | |
|---|---|---|---|---|---|---|
| | | *None* | *A.174* | *A.151* | *A.186* | *Z.6020* |
| Rheological properties, Monsanto oscillating disc rheometer | Torque (in lb, min.) | 10·5 | 10·0 | 11·0 | 11·5 | 12·5 |
| | Torque (in lb, max.) | 23·0 | 23·0 | 25·0 | 25·5 | 26·5 |
| | Torque change | 12·5 | 13·0 | 14·0 | 14·0 | 14·0 |
| | 90% cure time | 1·7 | 1·8 | 1·8 | 1·8 | 1·8 |
| Vulcanisate tensile properties (cure 10 min at 120 °C) | Tensile strength (MPa) | 5·75 | 4·24 | 5·84 | 5·94 | 4·83 |
| | Elongation at break (%) | 570 | 390 | 510 | 490 | 440 |
| | 100% modulus (MPa) | 1·45 | 1·94 | 2·01 | 2·08 | 1·69 |
| | 300% modulus (MPa) | 3·00 | 3·47 | 3·52 | 3·00 | 3·34 |
| | Young's modulus at 10% elongation (MPa) | 5·00 | 6·12 | 11·54 | 10·91 | 11·30 |

[a] Silane compositions:
A.174, γ-Methacryloxy-propyl trimethoxysilane
A.151, Vinyltriethoxysilane
A.186, β-(3,4-Epoxycyclohexyl)ethyl trimethoxysilane
} ex Union Carbide
Z.6020, N-β-(Aminoethyl)-γ-amino-propyl trimethoxysilane ex Dow Corning.

TABLE 7

COMPARISON OF SILICA-REINFORCED AND SILICA–CARBON FIBRE REINFORCED COMPOSITES

| Base mix: | |
|---|---|
| Silicone Gum 200 | 100 parts by weight |
| 2,4-Dichlorobenzoyl peroxide | 4 pphr |
| Silica R972 | As below, pphr |
| Grafil HTS, 3 mm | As below, pphr |

| | *Silica R972:* | *40* | *40* | *40* | *50* | *60* |
|---|---|---|---|---|---|---|
| | *Grafil HTS:* | *0* | *10* | *20* | *0* | *0* |
| Rheological properties, Monsanto oscillating disc rheometer | Torque (in lb, min.) | 10·0 | 12·0 | 15·8 | 17·4 | 24·0 |
| | Torque (in lb, max.) | 16·6 | 19·0 | 23·3 | 23·0 | 30·0 |
| | Torque change | 6·6 | 7·0 | 7·5 | 5·6 | 6·0 |
| | 90% cure time | 1·8 | 2·0 | 2·0 | 2·0 | 3·8 |
| Vulcanisate tensile properties (cure 10 min at 120 °C) | Tensile strength (MPa) | 7·68 | 6·45 | 5·45 | 5·41 | 4·03 |
| | Elongation at break (%) | 760 | 590 | 470 | 690 | 420 |
| | 100% modulus (MPa) | 0·76 | 1·17 | 2·10 | 0·87 | 1·42 |
| | 300% modulus (MPa) | 2·03 | 2·67 | 3·73 | 2·18 | 3·08 |
| | 500% modulus (MPa) | 3·96 | 4·80 | — | 3·83 | — |
| | Young's modulus at 10% elongation (MPa) | 2·06 | 5·38 | 10·10 | 2·54 | 4·20 |
| | Tear strength (kN/mm) | 17·7 | 7·3 | 9·5 | 14·5 | 7·4 |

### 5.8 Comparison of Glass Fibre with Carbon Fibre in a Composite

It had previously appeared that a poor fibre-to-matrix bond was obtained with carbon fibre, particularly after straining for 200–300% at higher fibre loadings. Consequently some composites containing 3 mm chopped glass fibre were prepared for comparison with the carbon fibre based materials to see if glass produced a better fibre–matrix bond. Test results (Table 8) were disappointing in that they showed somewhat poorer reinforcement (as evidenced by higher elongation at break and lower tensile modulus) for the glass fibre compared with the carbon fibre, even when the peroxide level had been increased with the glass fibre composite to improve the cure state. Evidence from a microphotograph (Fig. 17) showed that the glass fibre had broken down into shorter lengths similar to those found with the carbon fibre. No time was available to follow up this aspect of the project, though silane treatment of the glass fibre could well have improved the fibre–matrix bond.

## 6 CONCLUSIONS

1. Carbon fibre has been shown to increase the Young's modulus of a carbon fibre–silicone composite, though only to the extent predicted by the Guth–Gold equation. The original fibre length of 3 mm was found to have broken down to about an average of 0·1–0·2 mm owing to its brittleness.
2. Carbon fibre did not appear to interfere with the vulcanisation process in silicone rubbers.
3. Carbon fibre affected the normal stress–strain properties of silicone rubber. Up to a fibre content of 0·076 volume fraction (20 pphr) reinforcement of the matrix occurred over the full range of the stress/strain curve. Beyond the critical level of 0·076 VF the fibre-to-matrix bond had broken after an initial good reinforcement at low elongation, causing considerable flow of the matrix.
4. Carbon fibre did not improve the tear strength of silicone rubber.
5. Orientation of the carbon fibre had a significant effect on the properties of the composites. Alignment of the fibres was brought about by the shear stress exerted on the compound when passing through the nip of the two-roll mill during manufacture. This

TABLE 8

COMPOSITION OF GLASS FIBRE WITH CARBON FIBRE IN ELASTOMERIC COMPOSITES

| Base mix: | | |
|---|---|---|
| | Silicone Gum 200 | 100 parts by weight |
| | Silica R972 | 40 parts by weight |
| | 2,4-Dichlorobenzoyl peroxide | 2 or 4 parts |

| | | *Grafil HTS* | | | | *Glass fibre (3 mm)* | |
|---|---|---|---|---|---|---|---|
| | | *0* | *10* | *15* | *20* | *10* | *20* |
| | *Peroxide:* | *2* | *2* | *2* | *2* | *4* | *4* |
| Rheological properties, Monsanto oscillating disc rheometer | Torque (in lb, min.) | 11·0 | 15·0 | 16·0 | 17·0 | 10·6 | 11·8 |
| | Torque (in lb, max.) | 13·5 | 17·0 | 18·5 | 18·0 | 14·0 | 14·5 |
| | Torque change | 2·5 | 2·0 | 2·5 | 1·0 | 3·4 | 2·7 |
| | 90% cure time | 1·8 | 1·8 | 2·4 | 2·0 | — | — |
| Vulcanisate tensile properties (cure 10 min at 120 °C) | Tensile strength (MPa) | 2·94 | 2·11 | 1·57 | 1·27 | 3·14 | 2·08 |
| | Elongation at break (%) | 660 | 450 | 310 | 220 | 620 | 450 |
| | 100% modulus (MPa) | 0·60 | 1·02 | 1·01 | 1·06 | 0·97 | 1·19 |
| | 300% modulus (MPa) | 1·36 | 1·65 | 1·56 | — | 1·46 | 1·64 |
| | 500% modulus (MPa) | 2·36 | — | — | — | 2·39 | — |

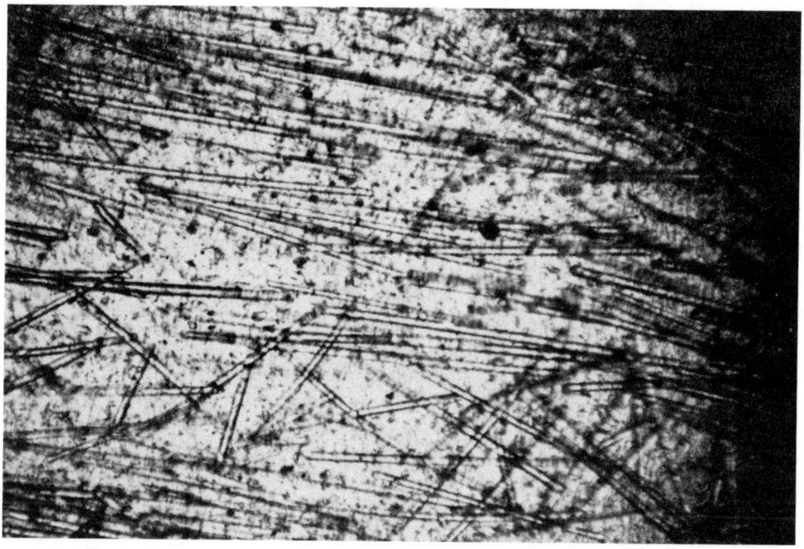

FIG. 17. Composite containing 0·04 VF (10 pphr) aligned glass fibre: longitudinal section through tensile bar, bar cut from sheet in machine direction (120 ×).

alignment was improved by prolonged milling of the unvulcanised rubber compound.

6. Chemical treatment of the carbon fibre surface using a silane coupling agent promoted the composite Young's modulus and hence improved the fibre-to-matrix adhesion at low strains. No real improvement, however, was apparent at high strains.
7. Carbon fibre was shown to be more beneficial in improving the stress–strain properties of the silicone rubber compared with the addition of an increased amount of reinforcing silica filler.
8. Glass fibre was not as satisfactory as carbon fibre for improving the properties of the silicone rubber.

## ACKNOWLEDGEMENT

The author wishes to acknowledge the permission of the Department of Industry to publish this paper, which is based on his own work and opinions and does not reflect the Department's views in any way.

## REFERENCES

1. *Journal of Composite Materials* from 1967 and *Composites* from 1969.
2. O'CONNOR, J. E. (1977) *Rubb. Chem. Technol.*, **50**, 945.
3. BLACKLEY, D. C. and PIKE, N. T. (1974) Presented at DKG Rubber Conference, September; later published in (a) *Kautsch. Gummi Kunstst.*, **29**(10), 1976, 607; (b) *Kautsch. Gummi Kunstst.*, **29**(11), 1976, 680; (c) *Kautsch. Gummi Kunstst.*, **30**(6), 1977, 367.
4. CAMPBELL, J. M. (1978) *Progress of Rubber Technology*, Vol. 41, Plastics and Rubber Institute, London, p. 43.
5. Bergougnan (Belgium), trade literature on 'Rubhose'.
6. GUTH, E., GIMHA, R. and GOLD, O. (1936) *Koll. Zh.*, **74**, 266.
7. ADAMS, D. K. (1951) USP 2 548 757.
8. DERRINGER, G. C. (1971) *J. Elastoplast.*, 3 October, p. 230.
9. ANTHOINE, G., ARNOLD, R., BOUSTANY, K. and CAMPBELL, J. (1975) *Eur. Rubb. J.*, July, p. 28.
10. MOGHE, S. R. (1976) *Rubb. Chem. Technol.*, **49**, 1160.
11. FOLDI, A. P. (1976) *Rubb. Chem. Technol.*, **49**, 379.
12. GILL, R. M. (1972) *Carbon Fibres in Composite Materials*, Iliffe, London, Section 8.7.1.
13. GENIN, G. (1964) *Plast. Inf.*, **15**(31), 1.
14. BERGSTROM, E. W. (1964) Report AD-69003, Rock Island Arsenal Laboratory, USA.
15. SIERON, J. K. (1963) *Rubb. World*, **148**(6), 50.
16. SIERON, J. K. (1966) *Rubb. Chem. Technol.*, **39**(4), 1141.
17. GOETTLER, L. A., LEIB, R. I. and LAMBRIGHT, A. J. (1979) *Rubb. Chem. Technol.*, **52**(4), 838.
18. SHEN, K. S. and RAINS, R. K. (1979) *Rubb. Chem. Technol.*, **52**(4), 764.
19. MANSON, J. A. and SPERLING, L. H. (1976) *Polymer Blends and Composites*, Plenum Press, New York, Section 12.2.
20. BROUTMAN, L. J. and KROCK, R. H. (1974) in *Composite Materials*, Academic Press, New York, Chaps 1–2.
21. NIELSEN, L. E. and CHEN, P. E. (1968) *J. Mater.*, **3**, 352.
22. UNION CARBIDE CORPORATION (1968) *Filler Treatment*, trade literature.
23. DOW CORNING CORPORATION (1970) *Silane Coupling Agents*, trade literature.

*Chapter 7*

# MISCELLANEOUS PRODUCTS

C. W. Evans

*Formerly Technical Director, Dunlop Limited, Gateshead, UK*

*SUMMARY*

*Continuing the theme of miscellaneous and unusual applications described in the 'Miscellaneous Products' chapter of Volume 1 of this series, a review is made of some unusual applications of rubber and rubber-like materials, not only as materials, but also the applications where some of their unique properties are used to advantage. In such a wide divergence of usage it is impossible to present the topics in any logical order or progression. Indeed, no attempt has been made to do this, but rather to present a balanced picture of materials, their usage and sufficient theory as is necessary to illustrate the particular topic.*

## 1 WEATHER FORECASTING

A range of weather balloons, identified by their weights in grams, from toy-like 10 g types to gigantic 4000 g models, are made with Neoprene latex. Neoprene is preferred to competitive latices because, in addition to the well-known environmental resistance and storage stability of the cured polymer, as a latex it is readily compounded for specific physical properties important in meteorological balloon design.

Typically, a balloon's initial wall thickness of approximately 0·076 mm decreases to about 0·013 mm at burst, which requires exceptional tensile strength of the spherically expanding film. Such thin cross-sections also call for a very homogeneous material to avoid flaws and weak spots.

80% of the work is done with the 600 g balloons. These rise on average to about 27 km in winter and 29 km in summer. Heavier, 1000 g units are substituted when there is severe weather and a balloon is needed that will withstand buffeting by high surface winds or when icing conditions may be encountered aloft. For higher altitude flights the larger balloons are used. A 1200 g balloon will ascend to 32–36 km. The little 10 g balloons are used for measuring the heights of clouds and the 30–100 g sizes for checking wind speeds aloft at low altitudes. The 100 g balloons are also used when upper air observations are needed up to 5 km for air pollution studies and forecasts.[a]

## 2 RADIATION PROTECTION

Radiation protection is afforded by lead-loaded dry box gloves. The lead is incorporated in the primary solvent dip formulation of Du Pont Neoprene used in making the gloves. It is reported that this process ensures maximum resistance to fluid permeation. A second, outer dip coating of Hypalon (Du Pont registered trademark) synthetic rubber enhances the gloves' resistance to abrasion, ozone or oxidising chemicals. These gloves are produced in two weights. The lighter is suitable for attenuation of soft gamma radiation (0·1 mm lead equivalent), while the heavyweight version (0·35 mm lead equivalent) is recommended for attenuation of higher radiation levels. Both are available in three port diameters, 20·3, 17·8 and 12·7 cm, and a standard length of 81·3 cm. Their primary application is in industrial nuclear facilities but they are also used in handling medical isotopes.[b]

## 3 ENERGY SAVING

Non-wrapped, cog-type transmission belts are claimed to be up to $4\frac{1}{2}$% more energy-efficient than conventional wrapped V-belts. The difference is attributed to the combination of moulded cogs and raw edge design. These features allow the belts to bend more easily around the sheaves,

[a] Information by courtesy of Du Pont de Nemours International and Kaysam Corporation of America.

[b] Information by courtesy of Du Pont de Nemours International and Norton Company, USA.

reducing the amount of energy wasted in bending the belt.

Multiplied by thousands of applications, the incremental reduction in electrical power consumption for the same work output is cited as converting into a significant potential reduction in crude oil requirements for industry. Apart from the direct energy conservation aspects of the belts, they are also strong and durable. Heat- and oil-resistant Neoprene construction throughout contributes to long service life. Users have reported sizeable maintenance and energy savings on all manner of industrial equipment, from hospital ventilation fans to lumber plant debarkers.[a]

## 4 VALVE STEM SEALS

Excessive oil consumption by an automobile engine is an obvious economic loss. But perhaps of equal concern in today's environment-conscious society is the route by which the oil escapes. For instance, serious problems can be created for the car owner if too much lubricant is entrained by exhaust gases. Phosphors in the oil could impair his catalytic converter's catalyst, increasing hydrocarbon emission in the short run and, ultimately, requiring expensive replacement of the anti-pollution device.

Automobile makers have worked at minimising oil loss down the valve guides of overhead valve engines by employing various sorts of valve stem seals. These designs fall into two principal types:

1. An umbrella-shaped shedder seal that is affixed to the valve stem and rides up and down with it.
2. A shaft seal that is affixed to the valve guide or engine boss and through which the valve stem reciprocates.

The former is referred to as a static seal and the latter as a dynamic or positive seal. Materials for each of these types of seals have included nitrile rubber, polyacrylates, nylon and Viton (Du Pont registered trademark) fluoroelastomer. Proprietary designs also embody metal casings, inserts, fastening rings, etc., particularly for the dynamic-type seals.

Exposure conditions call for hot oil resistance, low tension set and low modulus plus, in the case of the dynamic or positive-type seals, the good abrasion resistance characteristic of shaft seals. Although the exhaust

[a] Information by courtesy of Du Pont de Nemours International and Dayco Corporation, USA.

valves usually run 27 to 55 °C hotter than the intakes, the more critical seal application is on the intakes. This is because of the intake vacuum which can pull oil from the rocker arm gallery down the valve stems and into the cylinders.

Viton was used as early as 1958 in shedder-type valve seals on certain heavy-duty truck engines. Here they encountered temperatures exceeding 175 °C and performed satisfactorily for many years. More recently, several passenger car engines have been equipped with valve stem seals of Viton. Improved sealing at these points has been initiated, at least in part, in response to the demands for lower emission of hydrocarbons by local and national authorities. Major automobile firms throughout the world are currently actively engaged in valve stem seal programmes.

Despite the obvious design differences between valve stem seals and rotary shaft seals, heat-ageing requirements are sufficiently similar for existing data on one to be applicable on the other. A rotary shaft seal of Viton sealed satisfactorily for 6000 h against motor oil at 149 °C in a laboratory test run on a 4·45 cm shaft turning at 3600 rev/min. Competitive elastomers failed in 2000 h or less.

This raises the sometimes overlooked factor of seal replacement costs. These can far outweigh the price differential for the initial seals of Viton. Comparative tests at elevated temperatures show a typical Viton formulation to be softer than silicone but equivalent to polyacrylate and nitrile rubbers in hardness, equivalent to polyacrylate and superior to nitrile and silicone in tear resistance, and to have a lower 10% modulus than any of the other three competitive elastomers. (See Du Pont's *Engineering Properties Manual* on Viton.)

Although different automotive companies have different approaches to engine design and emissions control, and thus different requirements for valve stem seals, Viton offers a range of properties capable of meeting these requirements, and a service life expectation sufficient to warrant consideration on an economic basis as well.[a]

## 5 FOAM

### 5.1 History of Foam

Foam has been used in the bedding and furniture industry in the UK for more than 40 years, longer than in any other country in the world. Latex

[a] Information by courtesy of Du Pont de Nemours International.

foam was invented in the late 1920s at the Dunlop Research Centre near Birmingham, and the first batch was made there. It was produced by whisking up natural latex (the milky sap of the rubber tree) and liquid soap in an ordinary cake mixer, and baking it in an oven. This same principle, through considerably more refined, is still used today for some products.

The development of latex foam continues today, side by side with that of polyurethane foams. The manufacture of polyurethane foams began in the UK just after the Second World War. These chemically based foams were initially developed in Germany and the USA. Development in polyurethane foams has progressed very rapidly, and their horizons still appear to be virtually unlimited. Neomorphic and high resilient foams are the latest additions to this family of foams.

## 5.2 Foam Manufacture

### *5.2.1 Latex Foam*

Latex foam can be made in two ways; the first method was developed from the original Dunlop process, and the second was developed by Talalay. In the first method the liquid latex is compounded with processing chemicals and frothed, using a metered air supply in a specially designed machine. The density of the final foam can be adjusted by altering the latex/air ratio. A delayed-action gelling agent is put into the liquid foam so that the foam takes up the shape of the mould whilst still liquid and is subsequently set in this shape by the action of the gelling agent.

This process is used chiefly for making cavitied cushions. A series of numbered moulds is set on a conveyer, and the frother is programmed in advance to dispense the right amount of foam into each mould. This timing is calculated to within a tenth of a second to make the most economical use of the foam available. The foam is fed through a hose which is manually operated. The progress of the conveyor closes the mould, which is then carried on towards an oven, where the foam is vulcanised—the final treatment that gives it its resilience. At the other end of the conveyor the vulcanised foam is removed from the mould, passed through rollers where it is washed with jets of water, dried, and sent through for the next stage of its progress towards the end-user.

Cavitied cushions are used for cars and commercial vehicles, and also for furniture when quantity production justifies the cost of making a mould. The degree of support given by the finished cavitied cushion is dependent on three factors: the density of the foam, the overall shape and size of the unit, and the size and shape of the cavities—large cavities give a softer feel, so these are positioned in the central area, while smaller cavities in a greater depth of foam give the necessary support at the perimeter. The metal pillars that form the cavities are fixed in the lid of the mould, and the particular shape of individual products is produced by the outer contours of the mould in which they are made.

The same continuous frothing method of making latex can also be applied to the manufacture of sheet foam. The liquid foam is spread on a moving belt and the top is levelled by a doctor blade to the required thickness. It is cured and finished in exactly the same way as the moulded products, and the end-uses include toppings, padding, and hand-built cushions where only a few of a particular type are needed. Foam made in this way can also be given a textile backing for specialised uses.

The other method of making latex foam was developed in comparatively recent years. All the latex foam used in Dunlopillo mattresses and pillows is now made by this process, which involves large and expensive plant, but results in a more evenly textured foam. In this process the liquid latex is whisked into a foam using only a small proportion of air. A metered quantity is fed into a large fixed mould, only partially filling it. The lid of the mould is lowered and sealed, and the partly frothed foam is expanded by a vacuum action until it completely fills the mould. By a complicated built-in heat-exchange system the temperature is then rapidly lowered and the foam frozen into a solid block. Carbon dioxide gas is passed through it, and the temperature raised to 104 °C which vulcanises the foam. The mould is opened and the moulded unit removed and thoroughly washed and dried in the same way as in the earlier process.

Moulds for this method of manufacture yield a product whose look is characterised by the large number of slim, pencil-shaped holes that replace the traditional cavities. The holes are formed by prongs in the lid and base of the mould, positioned in such a way that they overlap when the mould is closed. These moulds are also used to make large blocks of latex foam for the furniture industry. Manufacturers whose capacity does not justify the expense of individual mould construction can buy latex foam in block form and fabricate their components from it.

### *5.2.2 Polyurethane Foam*

Polyurethane foam is made from by-products of the petrochemical industry. The most widely known polyurethane foams are polyether and polyester, but it is only the former that is used as a cushioning material in furniture and bedding. Polyester foams do play their part in the furniture industry, since they are used for fabric lamination, and polyester foam laminated fabrics are now used as a covering material.

Like latex foam, polyether can be made in block form or moulded. The process for making blocks is highly dramatic. The reactive starting materials together with suitable catalysts and surfactants are metered with extreme accuracy into a mixing chamber. The resulting liquid mixture is fed through a nozzle that moves back and forth across a moving paper trough. As the liquid is carried away from the nozzle, the chemical reaction causes frothing to begin, and the liquid rises rapidly to its required height. The heat generated by the chemical reaction is sufficient to 'cure' the foam without the application of external heat. As the continuous block of foam comes off the plant it is cut into predetermined lengths and conveyed to a storage area. Complete blocks of foam are converted to meet individual requirements by being cut, split and shaped into smaller components.

When polyether foam is moulded (the same principle of mass production applies as for latex foam) a metered quantity of liquid polyether is fed into the mould which, although shaped, is not constructed to produce cavities like a latex foam mould. There are two polyurethane foam moulding processes, known as 'hot' and 'cold' cure. In the former, the moulds containing the foam are passed through a heated oven. It is during this operation that the foam is 'cured'. In the latter process the chemicals used are such that the application of external heat is unnecessary, hence the name. Cold-cure cushions are recognisable by their greater degree of resilience, more like that of latex foam.

Polyurethane foam can be specially formulated to reduce its ignition characteristics, and this is of particular value in contract applications. Until recently the method adopted to achieve this end was by means of chemical additives, but these tend to reduce the load-bearing characteristics of the foam, with the result that it is necessary to increase the density of these special grades. It is also the case that when these foams are involved in a fire they generate considerably more smoke than standard foams.

Work is continually in progress to develop foams which have an inherently improved resistance to ignition from small sources, careless use of the latter being so often the cause of domestic fires. Equally, the search has been for foams whose spread of flame properties and smoke release are an improvement on these grades based solely on the traditional additive approach. One of the major successes in this work was the development of neomorphic foams, which are not primarily dependent upon chemical additives for their improved fire safety characteristics.

## 5.3 Barrier Foams

The Ferex foam range is based on a remarkable new cellular bi-elastomer which is incapable of supporting flame as tested by BS 476 Part 7 and which will protect other materials from a wide range of dangerous ignition sources. These materials have the added advantage of being soft, flexible and resilient and therefore adaptable to many applications. The range is specially recommended for use in contract upholstery or mattresses where resistance to high ignition levels is needed. Typical uses will be in bedding, furniture and equipment in high fire risk medical areas, such as mental hospitals, geriatric units and operating areas. Ferex can also be of assistance in situations where deliberate attempts at ignition may be made.[a]

### *5.3.1. Fire Behaviour*

When provoked with a source of ignition, even as large as 5000 kJ,[b] these new materials will not ignite, will not melt away from the heat source and will not produce burning drops. On contact with a flame, the material will partially erode leaving an inert structure that forms a char barrier over the substrate.

### *5.3.2 Typical Applications*

Some typical uses are as follows:

Type 1: back cushion wrap
Type 2: seat cushion wrap
Type 3: full-depth padding

They can be used as a wrap or interliner over other upholstery materials with consideration of the total construction, bearing in mind the size of fire source to be resisted or the degree of provocation expected. The thickness of the wrap affects its feel, ease of use and fire resistance, e.g. to resist a completely burning cover, depending on type, on a chair, an approximate minimum of 12 mm will be needed to provide complete protection of the interior.

[a] Foam information here and in Section 5.4 by courtesy of Dunlopillo Division of Dunlop Ltd.

[b] e.g. when tested by BS 476 Part 7 which uses energies considerably above this.

All grades can be used and handled as normal upholstery foams. Adhesive can be used in assembly, but the effect of fire on the adhesive surface joins must be examined. Traditional conversion equipment can be used and the material can be milled and profiled. Upholstery staples can be used for attachment and shears can be used for trimming and cut-outs.

The material is chemically inert and non-toxic in normal use. The cellular construction, flexibility and ease of working, coupled with its fire resistance, will enable its use in acoustic and architectural applications such as wall padding, duct lining, general noise control and insulation.

*5.3.3 Physical Form and Properties (DIF 191, 194 and 196)*

Black, cellular, bi-elastomer, available in sheet form or as cut parts.

| | |
|---|---|
| Size: | width 1500 mm (fixed)<br>length 2000 mm (variable) |
| Thickness: | minimum 5 mm<br>maximum 25 mm |
| Feel: | Soft, pliable and resilient, but varies with grade. |

| Type: | 1 | 2 | 3 |
|---|---|---|---|
| Density (kg/m$^3$) | 80 | 125 | — |
| Tensile strength (kPa) | 120 | 120 | 130 |
| Compression set at 50% | 30% | 25% | 10% |
| Hardness | 190N | 280N | — |
| Temperature range | −30 °C | −110 °C | continuous |

Thermal conductivity (W/m°K): 0·050–0·065 depending on grade.

*5.3.4 Tests to Relevant Specifications*

| | |
|---|---|
| FMVSS 302 | Self-extinguishing. |
| BS 4735/ASTM 1692-74 | Does not ignite. Glows at the edges in the immediate vicinity of the flame during the application of the flame only. Weight loss 3%. |
| BS 476 Part 7 | Class 1. |
| BS 5111 (XP2 Smoke Test) | Maximum obscuration 65%. Time to maximum obscuration: 5 min. |

| | |
|---|---|
| CAA8/FAA | Pass all grades. |
| UL94 | Pass rating 94 HF.1. |
| PSA/FR3 | Passes source 5. |

**5.4 Acoustic Absorption**

Protection against environmental noise can take two forms: firstly, devices that entirely prevent or greatly diminish the level of noise reaching the ear, e.g. ear plugs or defenders; and secondly, materials which absorb and contain the noise within a given area so that damaging noise levels are not transmitted to people working nearby. It is in this area that flexible polyurethane foams have a part to play. Flexible polyurethane foams possess extremely good sound absorption coefficients and do not suffer from the faults found in other acoustic media, in particular, mechanical resonances in the body of the material.

*5.4.1. The Advantages of Dunlop Process Polyurethane Foam*

The foam is light in weight, pleasant to handle and easy to bend and shape. It is continuous in construction, simple to fix, can be supplied in any thickness; above all, it is a very efficient absorber.

In order to appreciate further the advantages of foamed polyurethane, it may be useful to examine the properties of some of the other materials often used. Glass fibre, mineral wool and asbestos are materials that have been used as sound absorbers for many years. Unfortunately they have several disadvantages. Firstly, it is difficult to obtain thick layers of uniform density so that the absorption properties remain constant throughout the medium. Secondly, they are unpleasant to handle and difficult to attach to surfaces, needing a retaining medium which changes their acoustic performance. Thirdly, the materials shed fibres on equipment and personnel, they pack down under vibration and become detached from their retaining structures.

Foams used for sound absorption can be divided into two main types: (1) high-quality polyether and (2) neomorphic. Both these foams have their horizontal burning characteristics checked for quality control purposes by BS 4735, and show burnt extents of less than 80 mm and 40 mm respectively. Whilst both foams will resist ignition by ‘match-size’ sources, the above test does not assess the potential fire hazard in use. Also, dust and dirt contamination can affect these properties.

*5.4.2 Specially Formulated Polyether*

For acoustic use a specially formulated polyether has been developed. The special mechanical properties and flow resistance of this foam give excellent low-frequency absorption and good energy dissipation.

## 5.5 Neomorphic Foams

Neomorphic foams represent a step forward in safety standards. In comparison with conventional foams, neomorphic foams prove more difficult to ignite. Smoke generation and carbon monoxide levels are reduced. The thermal insulation ($K$ factor) is 0·25 Btu/h (ft$^2$, °F/in) all grades. Safe dry temperature: 16 h at 140 °F produces no appreciable deterioration. Low-temperature flexibility tested at −30 °C.

### *5.5.1. Application of Sheet Foam in Industry*

These foams are used extensively throughout industry for sound absorption and transmission loss, and are currently in use in the following areas:

Textile machinery
Tobacco machinery
Diesel engine enclosures
Oil drilling rigs
Bottling halls
Aircraft
Presses/stamping machines
Ultrasonic welders
Hi-fi cabinets
Noise measurement screens
Orchestral control screens
Hovercraft
Tractor cabs
Excavators
Commercial vehicles

### *5.5.2. Transmission Loss*

These foams, though primarily sound absorbers, do have an important part to play in improving transmission loss performance in conjunction with other materials. A barrier adhered closely to a noise source will tend to vibrate at the same critical frequencies and its performance can suffer. A substratum of specially formulated foam will lift the barrier material away from the surface and operate as a mass/spring damper system.

Special acoustic cladding for diesel engines has been developed by members of the staff of the Mechanical Department of the Admiralty Engineering Laboratory, UK, and processed by the design team of the Deltic Division of D. Napier & Son Ltd. The whole engine is enclosed, a few inches from the surface, by a substantial and heavy rigid skin which is completely isolated from the engine by a 2 in thick layer of open-cell polyurethane. An inner skin of resin-bonded glass fibre follows the general contours of the engine, filling in the cavities as much as possible, and is bonded to the foam. The outer surface of the foam is bonded to a heavy outer skin of lead sheet sandwiched between two layers of resin-bonded glass fibre. The material weighs some 6 lb/ft$^2$. Endurance running of the clad engine gave very useful reductions in noise level. The cladding is arranged in removable panels which allow easy access for maintenance.

### *5.5.3 Anechoic Chambers*

It is generally recognised that the most effective form of acoustic treatment consists of a series of wedges or pyramids of absorbent material completely covering the walls, floor and ceiling. The lower limit of the useful frequency range of an absorbent material for a free-field room is usually taken as the point at which the absorption coefficient exceeds 90%. In chambers where very low frequencies are to be measured, it is essential that very long, soft wedges are used. Anechoic chambers are used by the BBC, Southampton University and the Lockheed Corporation Research Departments.

In free-field rooms these wedges have distinct advantages. They do not create dust, they do not shed dangerous particles on people and equipment, and because of the foam's intercellular structure they retain their form indefinitely, giving consistent results over prolonged periods. Also, importantly, the frequency range over which free-field conditions can be obtained extends below 50 Hz.

Experience has shown that the most cost-effective wedge absorbers in recent years have a cross-section at the base of 8 in × 8 in (approx. 200 mm × 200 mm). But wedges can be supplied in any shape or size and manufactured to customers' requirements, including combined wedge blocks for ease of installation. In practice, a wedge absorber is not tapered over its entire length, the sides being left parallel near the base. Allowance can be made for special fixings and wedges can be cut to replace existing wedges in other applications.

## 6 SUSPENSION SYSTEMS

Hydragas is the second generation of interconnected suspension systems designed by Moulton Developments and manufactured by Dunlop exclusively for the British Leyland Motor Corporation.[a] It is an independent suspension system of great simplicity which offers a reconciliation of the conflicting requirements of ride and handling. The fluid interconnection between the front and rear units reduces the pitch frequency to give 'big car ride' while sufficient stiffness is maintained in roll for good handling without the use of anti-roll bars.

Features of Hydragas are:

Inert gas for springing; no creep.
Integral damper; no friction or seals to wear, hence car life durability.
Separate damping for pitch and bounce.
Car weight carried by water-based fluid under pressure.
Fluid hermetically contained, i.e. no pump, no glands.
Rear suspension incorporating a rubber-mounted cross tube forming a simple compliant sub-frame.
Small size of suspension unit avoids intrusion into car space.
Bonded bushes in load-carrying suspension pivots require no maintenance.

## 6.1 Design Principles

### *6.1.1 Introduction*

To improve ride in a car the springs must be softened, thereby reducing the resonant frequency of the sprung mass. On the other hand, to improve handling the springs must be stiff to counter roll when cornering. It is towards reconciling this conflict between ride and handling that the Hydragas interconnected suspension has been designed.

Hydragas suspension is essentially simple, comprising an integral spring and damper unit at each wheel station. The units use a sealed-for-life inert gas as a springing medium. The weight of the car is carried by water-based fluid under pressure and the units are interconnected front and rear, the whole system being hermetically sealed. The Hydragas damping system is designed to last the life of the car and the spring–damper unit requires no maintenance. Its unique arrangement of using a rolling diaphragm eliminates the friction and wear of sliding seals.

### *6.1.2 The Hydragas Unit*

The Hydragas unit is a nitrogen-filled spherical chamber welded on to

[a]Suspension systems information by courtesy of Dunlop Ltd Engineering Group, Suspensions Division.

the top of a displacer chamber. Between these two chambers are carefully shaped holes covered by rubber compression blocks to control the flow of fluid between the upper and lower chamber. This valve is a two-way device and provides the required suspension damping. The displacer chamber is hermetically sealed at its lower end by a load-bearing nylon-reinforced rubber diaphragm which rolls between the skirt and the piston in response to wheel movement.

A butyl liner on the diaphragm ensures hermetic sealing. Within the upper chamber is a butyl separator. The volume above the separator is charged with nitrogen at 250 lb/in$^2$ (1725 kN/m$^2$) at the time of manufacture of the unit. The gas provides the springing and is sealed for life. During the fluid-charging procedure, which is similar to the Hydrolastic, fluid fills the space between the separator and the rubber diaphragm. When the fluid reaches the nitrogen charging pressure it will compress the gas and lift the separator off the bottom of the sphere. There is negligible stiffness in the separator so the pressure in the gas will then be the same as that in the fluid. The car is trimmed in an unladen condition at about 330 lb/in$^2$ (2279 kN/m$^2$) to a fixed height.

### *6.1.3 Damping*

The rubber compression valve within the Hydragas unit provides the necessary bounce damping. Since there are no sliding seals, there is no friction and no wear. This allows the ‘car life’ damping that has been designed into Hydragas. The large quantity of water-based fluid which circulates avoids excessive heat build-up and maintains the constancy of damping under arduous conditions.

At low velocities the flow in the rebound (downwards) direction is confined to the main bleed orifice. In the bump direction (upwards) the fluid flows through the bump bleeds by lifting the small rubber flaps on the side of the compression block, as well as through the main bleed orifice. At high velocities the fluid will also flow through the main ports by progressively lifting the rubber compression blocks. The housing which surrounds the blocks controls the amount of precompression applied to the rubber and hence the flow characteristics.

Pitch damping occurs in the interconnection pipe system. In the Hydragas system the interconnection pipe is connected to the displacer (lower chamber) underneath the bounce damper and this configuration is ‘underport’. This allows complete separate and optimum damping in pitch and bounce. The Hydrolastic system, on the other hand, has the interconnection pipe connected to the chamber above the bounce valve (overport) and so in the pitch mode the damping can only be the sum of the damping in the interconnection pipe plus the bounce damping.

## 7 AIR SUSPENSION

Essentially, air suspension is an arrangement wherein the sprung mass is supported upon a number of nylon-reinforced rubber air springs each connected to a pressurised air supply via control valves. In consequence, the pressurised air springs effectively support the sprung load transmitted to them, in proportion to the static load capacity of the spring.

Subsequently, changes in payload are accommodated by proportionate changes in load pressure, by the provision of position-sensitive, directional control valves, arranged to interconnect (mechanically) the sprung and unsprung masses. These levelling or height-control valves comprise essential components in the practical control circuit and perform an important function in the overall behaviour of the system. Thus, in response to permanent changes in the chassis to ground height relationship (caused by changes in payload), the valves allow quantities of air to be admitted to, or exhausted from, the air springs according to demand, thereby maintaining the original proportionality in terms of air spring load reaction. These devices also allow the additional benefit of constant chassis to axle static height.

### 7.1 Theory

The static load capacity of air springs is expressed by:

Static load capacity = Static load pressure (gauge) × Effective area of spring

Thus

$$F = pA \tag{1}$$

Also, considering the basic equation for bounce frequency, we see that

$$f = 187{\cdot}7\sqrt{\left(\frac{S_A}{W_s}\right)}\ \text{cpm} \tag{2}$$

where $f$ = bounce frequency at axle, $S_A$ = stiffness at axle, and $W_s$ = sprung mass at axle. Also, vertical displacement

$$X_A = \frac{\text{Sprung mass}}{\text{Stiffness}}$$

Hence, bounce frequency may also be expressed by

$$f = 187\cdot7\sqrt{\left(\frac{1}{X_A}\right)}\text{ cpm} \tag{3}$$

from which

$$f \propto \sqrt{\left(\frac{\text{Stiffness}}{\text{Sprung mass}}\right)}$$

or

$$f \propto \sqrt{\left(\frac{1}{\text{Displacement}}\right)}$$

It is therefore evident that, in order to maintain suspension stiffness (and hence bounce frequency) at a value substantially proportional to payload carried, it is necessary to control the instantaneous value of the fraction $S_A/W_s$ in the equations above.

From consideration of the behaviour of loaded air springs, in conjunction with the gas laws, the stiffness characteristics may be expressed as follows:

(i) For static conditions:

$$\frac{\mathrm{d}F}{\mathrm{d}h}\text{ (isothermal)} = p\frac{\mathrm{d}A}{\mathrm{d}h} + A\frac{\mathrm{d}p}{\mathrm{d}h} \tag{4}$$

(iii) For dynamic conditions:

$$\frac{\mathrm{d}F}{\mathrm{d}h}\text{ (adiabatic)} = p\frac{\mathrm{d}A}{\mathrm{d}h} + An\frac{\mathrm{d}p}{\mathrm{d}h} \tag{5}$$

where $p$ = load pressure (gauge), $\mathrm{d}A/\mathrm{d}h$ = rate of change of effective area with vertical displacement, $A$ = effective area, $n$ = adiabatic constant (1·38), and $\mathrm{d}p/\mathrm{d}h$ = rate of change of pressure with vertical displacement.

It should also be noted that the $\mathrm{d}p/\mathrm{d}h$ term may alternatively be expressed by:

$$\frac{\mathrm{d}p}{\mathrm{d}h} = -\left(\frac{\mathrm{d}v}{\mathrm{d}h} - \frac{1}{v}\right)(p + p_a) \tag{6}$$

where $\mathrm{d}v/\mathrm{d}h$ = rate of change of volume with vertical displacement, $v$ = volume of static height, $p$ = load pressure (gauge), and $p_a$ = atmospheric pressure. Thus, rewriting eqn (5) above, we have:

$$\mathrm{d}F/\mathrm{d}h\ (\text{adiabatic}) = p - \frac{\mathrm{d}A}{\mathrm{d}h} + An\left[ - \left(\frac{\mathrm{d}v}{\mathrm{d}h} - \frac{1}{v}\right)(p+p_a)\right] \tag{7}$$

Also, if we substitute for

$$\frac{\mathrm{d}A}{\mathrm{d}h} = \text{constant } k_1 \text{ and } An\left(\frac{\mathrm{d}v}{\mathrm{d}h} - \frac{1}{v}\right) = \text{constant } k_2$$

which is applicable for particular static heights of the air spring, then eqn (6) may alternatively be expressed as

$$\frac{\mathrm{d}F}{\mathrm{d}h}(\text{adiabatic}) = pk_1 + (p+p_a)k_2 \tag{8}$$

indicating that air spring dynamic stiffness characteristics vary substantially in proportion to the instantaneous value of load pressure $p$. This statement applies particularly for conditions where the load pressure is very high and the introduction of the term for atmospheric pressure, $p_a$, has least effect.

**7.2 Design Considerations**

Referring to theory, we see that

$$\text{Bounce frequency } f \propto \sqrt{\left(\frac{1}{\text{Displacement}}\right)}$$

in the undamped system. In consequence, it can be shown that for a linear rate spring system (e.g. leaf, coil or solid rubber), low-frequency ride characteristics require the imposition of large static vertical displacements, and hence a substantial change in working spring height where large changes between laden and unladen conditions occur.

For example, in a linear rate spring arrangement having a natural frequency of 60 cpm, the initial vertical displacement (free height to loaded height) would amount to 9·78 in. This would produce an unacceptably large free height in the coil spring, and a prohibitive difference in laden to unladen chassis height for the majority of CV applications.

In the case of the air spring installation, however, different circumstances prevail. The load pressure sensitive control system permits regulation of the air spring load pressure in direct proportion to the instantaneous static loading. Therefore the air spring has the ability to absorb very large effective static displacements in relatively small overall heights, allowing the chassis to axle relationship to be maintained substantially constant for all loadings.

Hence, the air spring provides the designer with an efficiently packaged unit which, in conjunction with the control system, can be integrated into many suspension systems, allowing a wide choice of ride characteristics.

### 7.3 Ride vs Roll Stability

Referring to the theory, we see that

$$\text{bounce frequency } f = 187{\cdot}7\sqrt{\left(\frac{S_A}{W_s}\right)} \text{ cpm}$$

where $S_A$=stiffness at the axle and $W_s$=sprung mass at the axle. It can also be shown that, for the beam axle suspension, the roll stiffness is expressed by:

$$\text{Roll stiffness } RS_A = \frac{2S_A B_A^2}{\theta}$$

where $S_A$=stiffness at the axle, $B_A$=half spring base at the axle, and $\theta=57{\cdot}3°$. Hence we see that both equations above are expressed as functions of $S_A$ (stiffness at the axle) which, in turn, is proportional to the vertical stiffness of the spring unit. Therefore, by rearrangement of the equations, we see that

$$RS_A \propto f^2$$

for a particular suspension geometry and static loading, where $B_A$ and $W_s$ are constants. Consequently, the relationship between roll stability and ride characteristic involving functions of spring stiffness and suspension geometry is established, and must be considered in detail as the design of the suspension system evolves.

For many applications where low-frequency suspension characteristics are required, the available suspension geometry invariably demands the inclusion of additional roll-control devices in order to ensure adequate roll stiffness for the vehicle. These features may be accomplished in a number of ways, but the behaviour of any additional devices with respect to horizontal (drag) and vertical (bounce) reactions and driving and/or braking torque reaction must be taken into account.

In general, the range of natural frequencies available (for the simply supported mass) from fully laden air springs employed in the direct-acting mode is:

(i) for rolling diaphragms: 60–90 cpm
(ii) for bellows: 90–150 cpm

Also, variations in air spring stiffness characteristic may be obtained by the control of spring volume employed in the system. In this respect the rolling diaphragm may be employed with or without piston volume included, and also with or without extra volume included, whilst the bellows is employed only with or without extra volume included.

### 7.4 Lateral Location

Owing to the nature of the materials employed, and the method of construction of the air spring, it does not normally possess sufficient lateral rigidity to allow direct lateral/longitudinal location of suspension members. Therefore, the location of members and control of articulation invariably requires the provision of pan-hard rods, or similar mechanical links. Nevertheless, the flexibility offered by air suspension systems allows construction in single, tandem and multi-axle configurations (by interconnecting each axle spring) and axle load equalisation is obtained without the need for 'balance beams' or similar mechanical devices.

### 7.5 Damping

Suspension systems employing air springs are not self-damping to any significant degree, except for the small amount of frictional damping present at mechanical joints, etc. For instance, values for the ratio system damping/critical damping are typically in the order of 0·01 for bellows and 0·10 for rolling diaphragm air springs. Consequently, it is necessary to include separate vibration dampers in the system, normally one per wheel station. These invariably take the form of double-acting hydraulic telescopic units as supplied by specialist manufacturers.

## 8 WATER-LUBRICATED ROLLER BEARINGS

The bearing is specially formulated with Adiprene urethane rubber, and the shells are either brass to specification CZ110 in accordance with BS 2871, or laminated phenolic. Where non-corrosion properties are required, phenolic, being non-metallic, has considerable advantages and is therefore becoming increasingly popular. The internal surface of the shell is textured and, being chemically clean, gives a good key for the special bonding process.[a]

These bearings are completely interchangeable with standard rubber bearings. Adiprene has outstanding abrasion resistance, even in the presence of aggressive agents normally found in service such as sand and grit. This means less wear, less vibration, less noise and longer life. The coefficient of friction remains low irrespective of speed, and has been shown in recent tests to be lower than that for equivalent rubber bearings. Grease may be used during assembly, without any detriment to the bearing material.

This range of bearings is acceptable to Lloyd's Register of Shipping for use in A1 class vessels. Apart from their success in boats and hydrofoils, they are also widely used in industrial, mining and oil industry applications, particularly in pump engineering. Their advantages can be listed as follows:

1. The coefficient of friction remains low without appreciable alteration, irrespective of load or speed.
2. Grit and sludge in the lubricating water does not score or wear away the shaft appreciably as it may do with a metal bearing. Because of this, the life of a rubber bearing is usually longer than that of other types under comparable operating conditions.
3. In a rubber bearing a propeller shaft will operate without pounding, thus considerably reducing vibration, noise and the effects of slight misalignment. This is a particular advantage for high-speed shafts, where a bearing of suitable rubber and design may permit a shaft to rotate on a centre of gyration which may not coincide with the geometric centre, thus reducing the magnitude of the dynamic stresses set up by the shaft. For this reason, no useful purpose is served by using a rubber bearing alongside or in conjunction with external rigid bearings, particularly if the object is to damp out vibration.

[a] Information by courtesy of Dunlop Ltd, GRG Division.

# INDEX